Chemistry Research and Applications

Chemistry Research and Applications

A Closer Look at Carvacrol
Zak A. Cunningham (Editor)
2022. ISBN: 978-1-68507-627-6 (Softcover)
2022. ISBN: 978-1-68507-634-4 (eBook)

Fundamentals of Photocatalysis
Orva Auger (Editor)
2021. ISBN: 978-1-68507-374-9 (Softcover)
2021. ISBN: 978-1-68507-417-3 (eBook)

Polypropylene: Advances in Research and Applications
Théodore Marleau (Editor)
2021. ISBN: 978-1-68507-378-7 (Hardcover)
2021. ISBN: 978-1-68507-401-2 (eBook)

Boron: Advances in Research and Applications
Lynn Mcconnell (Editor)
2021. ISBN: 978-1-68507-231-5 (Hardcover)
2021. ISBN: 978-1-68507-259-9 (eBook)

Applications of Layered Double Hydroxides
Rajib Lochan Goswamee, PhD (Editor)
Pinky Saikia, PhD (Editor)
2021. ISBN: 978-1-68507-355-8 (Hardcover)
2021. ISBN: 978-1-68507-381-7 (eBook)

More information about this series can be found at
https://novapublishers.com/product-category/series/chemistry-research-and-applications/

Warren L. Gregoire
Editor

Polycyclic Aromatic Hydrocarbons

Sources, Exposure and Health Effects

Copyright © 2022 by Nova Science Publishers, Inc.

All rights reserved. No part of this book may be reproduced, stored in a retrieval system or transmitted in any form or by any means: electronic, electrostatic, magnetic, tape, mechanical photocopying, recording or otherwise without the written permission of the Publisher.

We have partnered with Copyright Clearance Center to make it easy for you to obtain permissions to reuse content from this publication. Simply navigate to this publication's page on Nova's website and locate the "Get Permission" button below the title description. This button is linked directly to the title's permission page on copyright.com. Alternatively, you can visit copyright.com and search by title, ISBN, or ISSN.

For further questions about using the service on copyright.com, please contact:
Copyright Clearance Center
Phone: +1-(978) 750-8400   Fax: +1-(978) 750-4470   E-mail: info@copyright.com.

NOTICE TO THE READER

The Publisher has taken reasonable care in the preparation of this book, but makes no expressed or implied warranty of any kind and assumes no responsibility for any errors or omissions. No liability is assumed for incidental or consequential damages in connection with or arising out of information contained in this book. The Publisher shall not be liable for any special, consequential, or exemplary damages resulting, in whole or in part, from the readers' use of, or reliance upon, this material. Any parts of this book based on government reports are so indicated and copyright is claimed for those parts to the extent applicable to compilations of such works.

Independent verification should be sought for any data, advice or recommendations contained in this book. In addition, no responsibility is assumed by the Publisher for any injury and/or damage to persons or property arising from any methods, products, instructions, ideas or otherwise contained in this publication.

This publication is designed to provide accurate and authoritative information with regard to the subject matter covered herein. It is sold with the clear understanding that the Publisher is not engaged in rendering legal or any other professional services. If legal or any other expert assistance is required, the services of a competent person should be sought. FROM A DECLARATION OF PARTICIPANTS JOINTLY ADOPTED BY A COMMITTEE OF THE AMERICAN BAR ASSOCIATION AND A COMMITTEE OF PUBLISHERS.

Additional color graphics may be available in the e-book version of this book.

Library of Congress Cataloging-in-Publication Data

ISBN: 978-1-68507-626-9

Published by Nova Science Publishers, Inc. † *New York*

Contents

Preface		vii
Chapter 1	Explaining the Environmental Fate of PAHs in Indoor and Outdoor Environments by the Use of Artificial Intelligence	1
	Svetlana Stanišić, Gordana Jovanović, Mirjana Perišić, Snježana Herceg Romanić, Tijana Milićević and Andreja Stojić	
Chapter 2	Polycyclic Aromatic Hydrocarbon Contamination in Sharks and Batoids (Chondrichthyes: Elasmobranchii) and Ensuing Ecological Concerns	37
	Natascha Wosnick, Mariana F. Martins, Gabriela A. V. Moog and Rachel Ann Hauser-Davis	
Chapter 3	Phytoremediation of Polycyclic Aromatic Hydrocarbons: From Agricultural Soils to Freshwater Resources	61
	Taylan Kösesakal	
Chapter 4	Fisheries Contamination Following the NE Brazil Oil Spill: A Case Study of PAHs Levels in Samples from the Fishery Industry	79
	Renato S. Carreira, Carlos G. Massone, Wellington Guedes, Ivy de Souza, Leanderson Coimbra, Otoniel Santana, Renato Fortes, Lilian Almeida and Arthur L. Scofield	

Chapter 5	**Associations between Biliary Polycyclic Aromatic Hydrocarbons, Biomorphometric Indices and Biliverdin as a Feeding Status Proxy in Mullet (*Mugil liza*) from a Chronically Contaminated Estuary in Southeastern Brazil** .... 95 Rachel Ann Hauser-Davis, Roberta Lyrio Santos Neves, Danielle Lopes Mendonça and Roberta Lourenço Ziolli	
Index	 111	

Preface

Polycyclic Aromatic Hydrocarbons (PAHs) are chemicals that occur naturally in coal, oil, and gasoline, and are also produced when wood, tobacco, and other materials are burned. Exposure to PAHs can lead to negative health consequences, so remediating their presence in the environment is significant. Chapter One uses machine learning and artificial intelligence to explain the behavior of PAHs in indoor and outdoor environments of a university building. Chapter Two discusses PAH exposure and associated effects in elasmobranchs. Chapter Three focuses on the potential effects of plants on the phytoremediation of agricultural soils and freshwater resources contaminated with PAHs. Chapter Four includes a case study of PAHs levels in samples from the fishery industry. Lastly, Chapter Five evaluates potential associations between biliary PAHs, biomorphometric indices and biliverdin applied as a feeding status proxy in mullet (Mugil liza) from a chronically contaminated estuary in Southeastern Brazil.

Chapter 1 - The environmental fate of polycyclic aromatic hydrocarbons (PAH) is complex, as they are emitted from natural and anthropogenic sources into the atmosphere and further distributed to soil and, to a lesser extent, ground and surface water. People nowadays spend more than 80% of their time indoors in developed countries and 85-90% in Europe, which results in elevated human exposure to poor air quality in homes, working environments, public buildings, or the means of transportation.

We are witnessing the emerging need for the application of machine learning (ML) and explainable artificial intelligence (XAI) for the capturing of the unique and defining factors and processes responsible for the complexity, non-linearity, interactivity, or cross-compartment interconnectivity of environmental phenomena. The investigations of mutual interdependencies and relations among organic pollutants and environmental factors that govern their distribution in various matrices are supported by the large availability of high-dimensional data and successfully justified in environmental science.

In this chapter, the authors have used ML and XAI to explain the behavior and environmental fate of PAH compounds, as constituents of $PM_{2.5}$ measured in indoor and outdoor environments of a university building. The authors have investigated complex non-linear interactions between outdoor and indoor PAH levels, and inorganic gaseous pollutants, trace elements, ions, radon, 31 meteorological parameters, the number of people in the amphitheater, and the time they spent indoors. To illustrate the potential of the presented methodology, the authors have chosen to characterize environmental conditions that shape the occurrence of benzo[b]fluoranthene in both environments and determine the importance and degree of impact that certain environmental factors exhibit on its levels.

Chapter 2 - Polycyclic Aromatic Hydrocarbons (PAHs) are ubiquitous organic compounds in the marine environment, originating mainly from anthropogenic sources. Due to their physico-chemical properties and ability to bioaccumulate in living organisms, PAHs are of great ecotoxicological concern, and their carcinogenic, and genotoxic properties make these compounds particularly harmful to exposed organisms. In this context, marine fauna assessments are paramount to estimate PAH environmental bioavailability, biochemical effects, as well as potential ecological effects. Evaluations in marine organisms, such as bony fish, are readily available. Studies concerning sharks and batoids (elasmobranchs), however, are still scarce, which is interesting as this group is an important part of marine trophic network interactions and highly threatened by anthropogenic activities, including chemical contamination. In this context, this chapter will discuss PAH exposure and their associated effects in elasmobranchs. Furthermore, ensuing ecological concerns and public health implications due to contaminated elasmobranch meat consumption shall also be reflected upon.

Chapter 3 - Polycyclic aromatic hydrocarbons (PAHs) are hazardous organic compounds with fused aromatic rings originating from incomplete combustion or pyrolysis of fossil fuels. The significant increase in organic pollutants such as petroleum hydrocarbons, PAHs, organic solvents and pesticides worldwide in the last few decades has brought about the necessity to develop new techniques for their removal from the natural environment. Bioremediation, or enhanced biodegradation, is an accepted approach to clean matrices such as soil, sediment, surface and groundwater contaminated with polycyclic aromatic hydrocarbons. A variation of bioremediation is phytoremediation, which is defined as the use of higher plants to remove pollution. PAHs can be adsorbed, accumulated, transported, volatilized, or biodegraded in a non-phytotoxic form by plants. Additionally, plants can degrade organic pollutants by stimulating the microbial community in the rhizosphere. PAHs are known as anthropogenic pollutants harmful to plants, animals and humans. Polycyclic aromatic hydrocarbons threaten human health, agricultural productivity and the environment due to their toxic, mutagenic, carcinogenic and/or persistent properties. The phytoremediation of agricultural soils and freshwater areas contaminated with polycyclic aromatic hydrocarbons is a promising environmentally friendly approach. This chapter focused on the potential effects of plants on the phytoremediation of agricultural soils and freshwater resources contaminated with polycyclic aromatic hydrocarbons. The removal of different PAH compounds through different plant species is discussed in plants at the physiological, biochemical and molecular levels and based on phytoremediation mechanisms.

Chapter 4 - From August 2019 to January 2020, more than 3,000 km of the north-eastern – and a part of the south-eastern – Brazilian coastline were periodically hit by crude oil residues, an accident considered the largest of its kind to ever occur in South Atlantic. The oil originated from a spill at sea, but the origin of the spill and those responsible are, to date, still unknown. Even considering the limitations in conducting field sampling and laboratory research imposed by the COVID-19 pandemic, environmental impacts in the aftermath of the oil spill have been demonstrated, including contamination of coastal (mangroves, estuaries, bays, beaches) and marine (coral reefs, seagrass meadows, rodolith beds) areas and associated fauna and flora, and the need to understand long-term impacts in the affected areas is paramount. Furthermore, socioeconomic impacts on traditional artisanal fisheries communities and tourist activities have already been mapped, revealing a scenario of increased economic, health and cultural vulnerabilities of the affected communities.

Chapter 5 - Polycyclic aromatic hydrocarbons (PAH) are rapidly metabolized and excreted by vertebrates, resulting in low PAH concentrations in tissues routinely used in biomonitoring efforts, such as muscle and liver. Therefore, alternatives have been applied in this regard, such as the assessment of biliary PAH concentrations and their metabolites. However, fish feeding status has been reported as affecting PAH metabolite concentrations and is still poorly understood. Furthermore, it is known that PAH may significantly affect piscine biomorphometric indices, although studies are still scarce in this regard. Therefore, the aim of this study was to evaluate potential associations between biliary polycyclic aromatic hydrocarbons (PAH), biomorphometric indices and biliverdin applied as a feeding status proxy in mullet (*Mugil liza*) from a chronically contaminated estuary in Southeastern Brazil.

Chapter 1

Explaining the Environmental Fate of PAHs in Indoor and Outdoor Environments by the Use of Artificial Intelligence

Svetlana Stanišić[1], Gordana Jovanović[1,2], Mirjana Perišić[1,2], Snježana Herceg Romanić[3], Tijana Milićević[2] and Andreja Stojić[1,2,*]

[1]Singidunum University, Belgrade, Serbia
[2]Institute of Physics Belgrade, National Institute of the Republic of Serbia, University of Belgrade, Belgrade, Serbia
[3]Institute for Medical Research and Occupational Health, Zagreb, Croatia

Abstract

The environmental fate of polycyclic aromatic hydrocarbons (PAH) is complex, as they are emitted from natural and anthropogenic sources into the atmosphere and further distributed to soil and, to a lesser extent, ground and surface water. People nowadays spend more than 80% of their time indoors in developed countries and 85-90% in Europe, which results in elevated human exposure to poor air quality in homes, working environments, public buildings, or the means of transportation.

We are witnessing the emerging need for the application of machine learning (ML) and explainable artificial intelligence (XAI) for the capturing of the unique and defining factors and processes responsible for the complexity, non-linearity, interactivity, or cross-compartment interconnectivity of environmental phenomena. The investigations of

---

* Corresponding Author's Email: andreja.stojic@ipb.ac.rs.

In: Polycyclic Aromatic Hydrocarbons
Editor: Warren L. Gregoire
ISBN: 978-1-68507-626-9
© 2022 Nova Science Publishers, Inc.

mutual interdependencies and relations among organic pollutants and environmental factors that govern their distribution in various matrices are supported by the large availability of high-dimensional data and successfully justified in environmental science.

In this chapter, we have used ML and XAI to explain the behavior and environmental fate of PAH compounds, as constituents of $PM_{2.5}$ measured in indoor and outdoor environments of a university building. We have investigated complex non-linear interactions between outdoor and indoor PAH levels, and inorganic gaseous pollutants, trace elements, ions, radon, 31 meteorological parameters, the number of people in the amphitheater, and the time they spent indoors. To illustrate the potential of the presented methodology, we have chosen to characterize environmental conditions that shape the occurrence of benzo[b]fluoranthene in both environments and determine the importance and degree of impact that certain environmental factors exhibit on its levels.

Keywords: PAHs, benzo[b]fluoranthene, air quality, PM-bound pollutants, artificial intelligence

The Environmental Fate of PAH Compounds

The Origin and Sources of PAHs

PAHs refer to a group of several hundred aromatic hydrocarbons with two to seven fused benzene rings bonded in either a linear, angular or clustered way. In general, PAHs are generated in the thermal decomposition of organic substances including high-temperature processes of incomplete combustion and/or long-term exposure of organic material to low temperatures up to 300°C. The lower the combustion temperatures are, the incomplete combustion process will release more PAHs. Additionally, under low temperatures, the generation of alkyl homologues is favored while the higher combustion temperatures are more associated with simple and condensed chemical structures. The group of low molecular weight compounds consists of two or three aromatic rings, while the PAH species with four or more benzene rings are assigned as high molecular weight PAHs (Zhao et al., 2020). According to the structure, PAHs can be divided into alternant compounds which are more planar and symmetrical. The compounds are derived by fusion of additional six-membered benzene rings and non-alternant species in which benzene rings can be connected by five-numbered structures, which are typically emitted from low-temperature combustion sources (Wick et al., 2011).

At room temperature, PAHs are colored, crystalline solids that have low solubility in water, high boiling points, and low vapour pressure. These properties are pronounced with the increase in molecular weight. Some features that can be associated with PAHs are persistence, fluorescence, light sensitivity, heat, and corrosion resistance and conductivity. Additionally, each PAH ring structure possesses unique UV absorbance spectra and this is used for the PAH identification.

These air pollutants are emitted from both natural and anthropogenic sources. Natural sources encompass forest and vegetation fires, oil seeps, volcanic eruptions, diagenetic processes, and exudates from trees. Some PAH species are produced from pigments of fungi, insects, and marine species in anaerobic conditions that can be found in soil or subaquatic sediments (Paulik et al., 2018). While the natural PAH emissions represent a significant source of these contaminants to the environment, anthropogenic sources remain dominant in urban and industrialized areas. Nowadays the anthropogenic fossil fuel combustion is the major source of these compounds, including traffic emissions, coal-gasification sites, and smokehouses, while other sources of PAHs include aluminum production plants, burning of wood, garbage, tobacco, plant material and refuse, use of lubricating oil and oil filters, municipal solid waste incineration, petroleum spills and discharge, coke production, as well as activities that include the use of coal tar, asphalt, creosote (wood preservation) and roofing tar. In urban locations where the traffic speed is low and gear changing is frequent, higher PAH emissions are registered, followed by higher PAH levels in the soil next to the road (Teixeira et al., 2015). Thereby, diesel vehicle emissions are associated with lighter molecular weight PAHs, while gasoline emissions tend to contain more heavy molecular weight PAH. The study of Slezakova et al., (2013) has shown that diesel vehicular emissions can represent the major source of PAHs in urban areas. Once emitted, they are distributed to the atmosphere and deposited on terrestrial and water surfaces.

Taking into consideration the potential for toxicity and carcinogenicity, prevalence in hazardous waste, and persistence in environmental conditions, the U.S. EPA has listed 16 PAH compounds on the Priority Pollutant List created under the Clean Water Act. These PAH species were prioritized not only due to their toxicity but also due to their frequency of occurrence and potential for human exposure. For the purpose of certain location risk assessment, concentrations of these species, often referred to as the "priority

PAHs," are generally monitored: naphthalene (NAP), acenaphthylene (ACY), acenaphthene (ACE), fluorene (FLU), phenanthrene (PHEN), anthracene (ANTH), fluoranthene (FLTH), pyrene (PYR), benzo[a]anthracene (B[a]A), chrysene (CHRY), benzo[b]fluoranthene (B[b]F), benzo[k]fluoranthene (B[k]F), benzo[a]pyrene (B[a]P), benzo[g,h,i]perylene (B[ghi]P), indeno[1,2,3-c,d]pyrene (IND), and dibenz[a,h]anthracene (D[ah]A) (Samburova et al., 2017). Generally, phenanthrene, fluoranthene, and pyrene are dominating the urban air.

Human Exposure to PAH Compounds and Health Effects

As widely distributed environmental contaminants, PAHs were among the first atmospheric pollutants that were confirmed to be toxic, mutagenic, and carcinogenic species with detrimental effects on human health and living organisms. Due to their ubiquitous occurrence in air, water, terrestrial, and biological systems and physicochemical properties which support their distribution to all environmental compartments, resistance to biodegradation, a tendency for bioaccumulation, and carcinogenic potential, PAHs have gathered significant environmental concern. Due to their lipophilic nature, these compounds readily penetrate the skin, as well as cellular membranes. Among PAH contaminants, benzo(a)pyrene is considered the most carcinogenic and toxic, followed by benzo(a)anthracene and dibenz(ah) anthracene.

Other health-relevant possible/probable carcinogenic PAHs (IARC, 2012) include chrysene (Chy), benzo[b]fluoranthene (B[b]F), benzo[k]fluoranthene (B[k]F), and indeno[1,2,3-cd]pyrene (I[cd]P). Generally, with the increase in molecular weight, PAH carcinogenicity is more pronounced, while acute toxicity is reduced. Therefore, low molecular weight PAHs are considered directly toxic, while heavy molecular weight PAH compounds are considered genotoxic or capable of causing damage to DNA (Kim et al., 2013). Although the lighter-weight compounds are in some way less hazardous, their reactions with ozone and gaseous oxides result in the formation of highly toxic substances such as diones, nitro- and dinitro-PAHs, and sulfuric acids. On the other hand, the carcinogenic potential of certain heavy-weight PAHs depends on the compound and its metabolic route. Namely, as a result of PAH metabolic transformations in the human body, a complex mixture of quinones, quinines, cis- and trans-dihydrodiols, phenols, epoxides, and other oxidized metabolites are generated and these metabolites are able to covalently bind to nucleic acids and induce strand breaks and DNA damage, which results in genetic mutation (Bansal and Kim, 2015).

Human exposure to PAHs is related to cigarette smoking, inhalation of polluted air, dermal exposure from occupational or non-occupational settings, eating agricultural products grown in the PAH contaminated soil, and even more by intake of food which has been grilled, smoked, or roasted.

Nevertheless, assessing human health risk can be very complex due to the carcinogenic or detrimental potential of PAHs in combination. Therefore, this procedure is simplified by the introduction of the toxicity equivalency factor (TEF). TEFs are assigned to different species based on their relative toxicity compared to known human carcinogen B[a]P, chosen to be a reference compound, and then the concentrations of carcinogenic PAHs are converted to an equivalent concentration of B[a]P mixtures (Hussar et al., 2012).

The extent of exposure, the contaminant type, and its concentration are the main predictors of adverse health effects. The factors that should be also considered are related to the route of exposure and pre-existing health conditions. The majority of studies have been investigated human exposure to a mixture of PAH compounds, and moreover to the mixture of air pollutants including non-PAH potentially carcinogenic contaminants, while the studies on animals have considered the effects of the exposure to higher levels of individual contaminant. The latter have shown that the exposure to benzo(a)anthracene, benzo(a)pyrene, and naphthalene has been linked to embryotic effects in experimental animals, while the ingestion of benzo(a)pyrene during pregnancy can lead to low birth weight and birth defects (Vignet et al., 2014).

As regards acute health effects in the human population, it has been recognized that PAHs can provoke an exacerbation of asthma and thrombosis in people with atherosclerotic formations. Furthermore, occupational PAH exposure during coke production, bituminous product use during roofing or oil refining can lead to eye irritation and gastrointestinal symptoms of intoxication. Additionally, anthracene and benzo(a)pyrene are direct skin irritants and sensitizers, responsible for skin irritations and allergies in people who are exposed to their high concentrations (Friesen et al., 2010). As regards chronic health effects, it has been already mentioned that long-term exposure to some PAH species is associated with the binding of electrophilic PAH metabolites to DNA, gene mutation, DNA damage, and consequentially higher incidence of skin, bladder, respiratory, and gastrointestinal cancer. Besides, certain PAHs have the potential to interfere with hormones, which increases the chances of hormonal disruption and immune and reproductive system failure. In compliance with the findings of animal studies, it has been evidenced that exposure to PAH pollution during pregnancy might be related to adverse birth outcomes, low birth weight, premature delivery, low IQ, behavioral problems, childhood asthma, and delayed child development (Kim et al., 2013). Finally, it might be worth considering some indirect adverse health effects of PAH pollution. Namely, the study of Parajuli et al., (2017) has concluded that pollution-induced shifts in natural soil bacterial population which is registered in PAH-polluted areas can contribute to the prevalence of chronic diseases in the local population. These findings have been explained by the fact that he human microbiome, which has been shown to play a significant role in the immune system, is affected by the environmental microbiome.

In a number of studies, the PAH exposure assessment was conducted by the determination of certain pyrene metabolites in the urine. The level of this exposure biomarker in the urine can be two times higher in smoking compared to the non-smoking population.

Dietary Exposure to PAH Compounds

The current scientific evidence and knowledge suggest that dietary intake of PAHs is the major route of human exposure for the non-smoking population in particular. Besides, over the last three decades, human exposure to PAHs has been increased not only in developing countries where pollutant emissions are on the rise but also in developed countries due to their wide spread in food chain, modern lifestyle, and heavy reliance on fast food. In compliance with the aforementioned, dietary intake of PAHs is suggested to be one of the major factors contributing to skin and lung cancer, but also metabolic disorders and non-genotoxic diseases including diabetes mellitus and cardiovascular disorders.

The common nutrition sources of PAH refer to a wide variety of foods, including raw fruits and vegetables, smoked and grilled meat, refined fats and oils, fatty fish and seafood, etc. (Zelinkova and Wenzl, 2015).

The food that contains the highest PAH concentrations is the one processed by grilling, roasting, smoking, and frying. The amount of PAH in food depends on the heat source, the distance of heating, the design of the food device, and the type of fuel being used. Previous research has shown that PAH formation in the open-flame grilled meat was reduced when pre-heating with steam and microwave or wrapping with aluminum and banana leaf were applied, with the latter being more effective. The removal of smoke during meat grilling can lead to a 74% reduction in PAH content (Lee et al., 2016). In addition to this, PAH formation is prevented by the consumption of lean meat and fish, by the avoidance of open flame and direct food contact of meat and flame during barbequing, by cooking at lower temperatures, by using electric or gas meat broilers over charcoal, and by using the acid-base marinade as lemon juice before grilling of meat.

The process of food smoking results in PAH emissions during the incomplete combustion of wood. The use of poplar and hickory can result in a decrease of up to 55% in the PAH food contents compared to the commonly used beech wood (Hitzel et al., 2013). In developed countries, the majority of food products are treated with liquid smokes, and this alternative to the smoking process doesn't cause air pollution and allows for better control over PAH concentrations in the final product (Varlet et al., 2010). The possible approach to PAH reduction in smoked food products includes the immediate use of low-density polyethylene (LDPE) layer over smoked meat. This procedure has been proved to reduce the initial PAH level in food by more than 50% because the surface of the smoked/grilled foods contains most of the PAHs which can diffuse to packaging film of similar polarity immediately after wrapping. Moreover, polytetrafluoroethylene (PET) has been also shown to be useful for PAH concentration reduction in refined seed oil (Bansal and Kim, 2015).

Fruits and vegetables which are grown in the vicinity of industrial sources or areas with dense traffic contain higher amounts of PAH, but rarely more than 5 µg kg^{-1}. Thereby, it has been evidenced that leafy vegetables are more contaminated than stem vegetables such as cucumber, eggplant, and tomato, due to their larger, waxy, and cuticle surface area, which is vulnerable to the PAH deposition. Similarly, the root vegetables are more prone to PAH uptake from contaminated soil. Trace levels of phenanthrene, fluoranthene, and pyrene have been found in every raw fruit and vegetable (Paris et al., 2018). The concentrations of PAHs are lower in raw food grown in the areas not affected by volcanoes, forest fires, traffic, and industrial emissions. Additionally, washing fruits and vegetables with oxidizing agents can be useful for lowering the PAH content.

Seafood also contains certain PAH amounts, depending on the lifetime accumulation and water contamination, disposal of industrial effluents near coastal waters, proximity of oils spills, and use of creosote-treated wood for mussel cultivation (Gohlke et al., 2011; Rotkin-Ellman et al., 2012). Zhao et al., (2014) have shown that PAH concentrations can reach up to 513 µg kg^{-1}

in different tissues of bighead carp and silver carp. Fats and oils are also significant dietary sources of PAHs (Hao, Li and Yao, 2016). Furthermore, chocolate sweets contain PAH due to the drying, roasting, winnowing, blending, and fermenting of the cocoa seeds (Lowor et al., 2012).

Soil as the Ultimate Sink of PAHs

Although they have been investigated as air pollutants, for PAHs soil represents the ultimate sink. Levels of PAH compounds in a particular soil can typically range from 1 to 10 µg kg^{-1} depending on the proximity of the emission sources. Nevertheless, the total concentrations of 16 US EPA priority PAHs (ΣPAHs) can reach much higher levels. As shown in the study of Zhu et al., (2019), the average ΣPAHs concentrations were 274 ng m^{-3} in the air, 255 µg kg^{-1} in the soil, and 15 µg kg^{-1} in vegetation at a measurement site located in the economic and industrial center The Yangtze River Delta.

Most PAHs are deposited from the atmosphere in the soil and leached from overlying horizons to greater depth and lower aggregate surfaces. Due to their low solubility in water, they are found in low concentrations in soil water and more often are bound to soil particles that are dispersed in soil solution (Wilcke, 2000).

In the environment, PAHs undergo adsorption on soil particles, absorption by plant roots or animal ingestion, volatilization from soil, plant or water surfaces, leaching or translocation laterally or downward through the soil, photo-oxidation, chemical and microbial degradation, alteration by chemical processes such as oxidation-reduction reactions, adsorption through interaction with soil and sediments and diffusion into soil micropores where they become unavailable for microbial degradation (Johnsen, Wick and Harms, 2005). The processes of transfer, degradation, and sequestration occur through a variety of mechanisms and are highly dependent on PAH molecular weight, structure, water solubility, and vapour pressure. For instance, their persistence increases with molecular weight (Yukhimets et al., 2019).

The levels of PAHs are highest at the specifically contaminated sites, such as gasworks sites due to waste materials, coke ovens, petroleum refineries, and wood conservation plants, followed by urban soils, where the contribution of industrial and traffic emissions and emissions from fossil fuel burning for heating operations are significant. Lower PAH concentrations were registered in permanent grassland, forest, and arable soils. The PAH mobility that occurs in the soil doesn't correspond with molecular weight as it would be expected due to the fact that high molecular weight PAH species are less water-soluble. Sorption and desorption are considered to be the main processes for PAH transport in the soil, which suggests that high molecular weight PAHs are mostly transported as particle-adsorbed. The significant share of organic pollutants that remain in prolonged contact with soil is adsorbed on soil particles in the process of contaminant sequestration which makes them also less immobilized and less available for biodegradation (Haritash and Kaushik, 2009). Thus, the transport of PAHs in the soil is described by a model which includes the solid, the dissolved and the particle-bound phase.

In the organic horizons of forests and urban soils, individual PAH concentrations can reach several 100 $\mu g\ kg^{-1}$. The presence of different compounds depends on the climate, but in temperate soils, the most dominant species are benzofluoranthenes, chrysene, and fluoranthene (Wilcke, 2000). Furthermore, meteorological features exhibit a strong impact on PAH dynamic patterns in soil. For instance, moisture and temperature affect PAH decomposition and volatilization.

PAH levels are assumed to be 10 times higher than the concentrations that were present prior to industrialization, which suggests the share of anthropogenic contribution over the last century. In most remote areas like the Arctic, benzo(a)pyrene levels are in the range of those from the preindustrial era.

Nowadays, various techniques have been developed to treat PAH-contaminated soil, including soil washing, chemical oxidation, electrokinetic, and phytoremediation. The study of Gitipour et al., (2018) reported that the surfactant-aided washing process had a 90% efficiency, while compost-amended phytoremediation removed from 58 to 99% of pyrene from the soil over the 90-day period. Additionally, the most efficient treatment procedure was shown to be chemical oxidation, while the electrokinetic treatment has been proved successful in removing specific PAH contaminants.

The major PAH degradation natural process is related to microbial degradation generally catalyzed by enzymatic systems of microorganisms. Microbial degradation refers to PAH transformation into less complex and less hazardous/non-hazardous compounds and depends on weather conditions which can be aerobic and anaerobic. The final products of the degradation process are inorganic minerals, water, carbon dioxide, and methane. In addition to the fact that microbial degradation is a natural process, contaminated locations can be remediated by microbial manipulations and for this purpose, it is very important to understand the environmental fate of specific PAH compounds (Zhang et al., 2006). The rate of biodegradation depends on: 1) the environmental conditions, including pH, temperature, nutrients, metals, moisture, and oxygen presence, 2) microbial species such as algae, bacteria, and fungi and related factors including their population, degree of acclimation, accessibility of nutrients, cellular transport properties, and chemical partitioning in the growth medium, and 3) nature and chemical structure of the PAH compound being degraded. For instance, previous research has reported anaerobic degradation of two- and three-ring PAHs, but it has not been shown for PAHs containing more than three rings. Furthermore, higher molecular weight PAHs have been shown to be more resistant to biodegradation compared to low molecular weight PAHs. The studies have been performed *in situ* (contaminated sites), *ex situ* (bioreactors), or in laboratory settings with soil samples being PAH spiked. As shown, some microorganisms are capable of using PAHs as a source of carbon and energy, transforming the contaminants into nontoxic products (Ghosal et al., 2016).

Some "dead-end products" of many biological and chemical degradation pathways such as oxygenated PAHs are also known for their toxicity and carcinogenicity, which can be particularly important in cases where microflora is used for bioremediation purposes. Namely, it has been reported that badly designed bioremediation soil treatments resulted in the microbial formation of new, more toxic, more persistent, and more water-soluble and thus, mobile co-contaminants than those initially detected (Lundstedt et al., 2007). Finally, the bioremediation process might be limited by the supply of nutrients for bacterial population, non-optimal conditions related to temperature, pH, oxygen presence, and salt content, as well as lack of bacterial species that can degrade PAH compounds or low PAH availability due to their physicochemical features (Wick et al., 2011). Furthermore, the rate of PAH degradation in the soil can be also decreased due to contaminant adsorption which is dependent on soil features including cation exchange capacity, micropore volume, soil texture, and surface area. The biodegradation PAH rate and remediation of soil can be enhanced by increasing PAH bioavailability by plant establishment, increasing metabolic potential of the bacterial population through the addition of specific bacterial strains, or supplementation of contaminated sites with light oil, straw, fertilizer, manure, and compost material which serve to improve soil texture, oxygen transfer, and provide nutrients for the bacterial population. In general, without supplementation, only three-ring aromatics can be degraded.

The uptake of PAHs by plants is not significant because the plants are unable to transport hydrophobic substances and most of these species detected in plant tissue originate from atmospheric deposition, although the uptake by above-ground parts of the plant can also contribute. Vegetation in rural regions can contain from 50 to 80 μg kg^{-1} of PAHs, while the urban vegetation can have up to 10 times more PAH levels. Nevertheless, living beings in the soil accumulate considerable amounts of PAHs in a short period either by soil ingestion directly or by plant ingestion indirectly.

PAHs in the Atmosphere

The levels of PAHs in the atmosphere are dependent on season, meteorological conditions, time of the day, measurement site, but also on certain factors that in addition to weather conditions have an impact on atmospheric chemistry, dry or wet deposition, and finally, PAH half-lives and their interactions with other pollutants including ozone, SO_2, SO_3, NO_x, or OH radicals (Lee, 2003). In the cold season, PAH concentrations are higher due to increased fossil fuel burning for heating, reduced thermal and photo-decomposition, and less atmospheric mixing associated with the lower planetary boundary layer.

In the atmosphere, PAHs are found in the vapor phase or adsorbed in particulate matter. The partitioning of PAHs between particles and gas is mainly dependent on the atmospheric conditions and nature of the contaminants and aerosol including PM, soot, dust, fly ash, pyrogenic metal oxides, pollens, etc. At ambient temperatures, most atmospheric PAHs are particle-bound. The atmospheric transformations and degradations of PAHs in the atmosphere are reduced or inhibited when they are adsorbed to particles. Thereby, the lower molecular weight contaminants are more volatile and found in the gas phase, while particle-bound species are mostly those of higher molecular weight. In the cold season, particulate phase-bound PAHs are dominant, while in the warm season, gas-phase PAHs are often registered (Vuong et al., 2020). The sorption to particulate matter which increases with a decline in PAH volatility enables long-range transport of these species. As well as for the particulate matter, atmospheric precipitations appear to have a particularly scavenging efficiency for atmospheric PAHs (Gaga and Ari, 2019).

According to the report of NAEI (Brown et al., 2012), out of 621 tons of 16 priority PAHs being emitted from anthropogenic sources, 3.23 tons referred to benzo(a)pyrene, while the contribution of natural sources to total benzo(a)pyrene was negligibly lower (2.88 tons). Thereby, the absolutely highest contribution of all anthropogenic sources was attributed to fossil fuel combustion for heating purposes. While the PAH levels have declined over the previous decades due to the prevention measures and environmental protection policies, the air quality in developing countries where energy production relies on biomass and coal combustion remains low (Zhu et al., 2019). In areas where industrial emissions represent the major PAH sources, registered contaminant levels exhibit no seasonal variations. However, in the residential areas which are dominated by coal and wood burning for heating, higher PAH concentrations are registered in winter and the seasonal concentration dynamic is pronounced (Gune et al., 2019). Among the factors that have an impact on outdoor PAH concentrations, the study of Slezakova et al., (2013) has emphasized the significance of other gaseous air pollutants and solar radiation, while ozone, temperature, relative humidity, and wind speed were recognized to be negatively correlated with PAH concentrations. The study of Elorduy et al., (2016) emphasized that out of meteorological parameters, wind speed, temperature, and atmospheric pressure were recognized as significant for governing distribution of PAHs in the air whereas the influence of relative humidity appeared to be negligible.

The $PM_{2.5}$-bound PAH Behavior in Indoor and Outdoor Environments: the Explainable Prediction of Benzo(b)fluoranthene Levels

People nowadays spend more than 80% of their time indoors in developed countries and 85-90% in Europe, which results in elevated human exposure to poor air quality in homes, working environment, public buildings, or the means of transportation (Ferguson et al., 2020; González-Martín et al., 2020). However, literature sources on the occurrence and behavior of air pollutants in indoor environments are still scarce, although the levels of indoor PAH compounds can exceed their concentrations in the outdoor environment. For instance, the study of Cao et al., (2019) aimed at investigating spatial variations and corresponding health risks of PAHs included composite settled dust samples collected from four types of microenvironments including offices, hotels, dormitories, and kindergartens in Beijing. As shown, the total concentrations of explored PAH species ranged from 388 μg kg^{-1} (kindergarten dust) to 8140 μg kg^{-1} (hotel dust). Indoor air quality clearly depends on outdoor contaminants intrusion and on the activity of endogenous sources such as emissions from building materials, cleaning and disinfection products, cooking and heating, and an individual's activities (González-Martín et al., 2020).

The previous literature findings regarding benzo(b)fluoranthene (B[b]F) indoor levels (González-Martín et al., 2020; Oliveira et al., 2019) reported highly variable data which can be attributed to the fact that different sampling periods and measurement sites were chosen, as well as to the fact that different methodologies were applied. Additionally, the features of a certain measurement location including surrounding emission sources, topography, and meteorological conditions, make a regional and global comparison of data less feasible. According to the study of Pereira et al. (2019), B[b]F was was among the most abundant PAH in outdoor (0.04 ng m^{-3}) and indoor (0.06 ng m^{-3}) school environments and based on the B[b]F/B[k]F diagnostic ratio, its' origin was recognized to be associated with light vehicular diesel emissions. Unlike in Europe or North America, PAH imposes a significant burden in the school environment in Asia (Oliveira et al., 2019), where average indoor and outdoor PM$_{2.5}$-bound B[b]F levels were 25.3 and 27.1 ng m^{-3}, and 0.95 and 1.08 ng m^{-3}, during the cold and non-heating season, respectively (Zhang et al., 2020). Extremely high levels of PM-bound B[b]F reaching 174.7 ng m^{-3} were found in indoor water pipe cafés of Ardabil city, Iran (Rostami et al., 2021).

The emerging need for the application of (explainable) artificial intelligence (AI) and machine learning (ML) methods is supported by the large availability of high-dimensional data (Blair et al., 2019; Gilbert et al., 2018) and successfully justified in environmental science for the investigations of mutual interdependencies and relations among organic pollutants and environmental factors that govern their distribution in various matrices (Jovanović et al., 2019; Stojić et al., 2019; Stanišić et al., 2021). In this study, we used eXtreme Gradient Boosting (XGBoost) to study the behavior of B[b]F in the indoor and outdoor environments of a university building. We also employed SHapley Additive exPlanations (SHAP), an additive feature attribution method that provides a *post-hoc* explanation of the ML models' overall behavior in the form of feature contributions, which is more aligned with human intuition (Chen et al., 2021; Stojić et al., 2019). The methods were applied to: 1) explain complex non-linear interactions between outdoor and indoor B[b]F levels, and inorganic gaseous pollutants, trace elements, ions, radon, 31 meteorological parameters, the number of people in the amphitheater, and the time they spent indoor, 2) characterize environmental conditions that shape B[b]F occurrence in both environments, and 3) determine the importance and degree of impact that certain environmental factor exhibit on B[b]F levels.

Methodology

Measurement Campaign

During the three-month study campaign (March 1st – May 31st), the sampling of organic and inorganic criteria air pollutants was performed in indoor and outdoor measurement sites of the building of Singidunum University (44°45'33.8"N, 20°29'47.6"E), situated in the urban area of Belgrade, the capital of Serbia. The University building is placed within large residential areas, which contain households with individual fireboxes using coal and wood; heating plants and fuel oil heating plant using natural gas and crude oil around 800 m to the W and SW; several small-scale chemical plants and food factories located up to 600 m in the NW and S direction; a boulevard with public transport and moderate vehicle flow approximately 250 m in the SW direction and a road with intense traffic about 500 m away in the W–NW direction.

The outdoor sampling of $PM_{2.5}$, air and ambient air temperature, outdoor relative humidity, outdoor air pressure, wind and rain characteristics was carried out at the rooftop of the University building, around 10 m above the ground while additional data on 24 meteorological parameters were obtained from Global Data Assimilation System (GDAS1). The indoor air and $PM_{2.5}$ sampling were performed at a height of 6 m and 2 m from the floor, respectively, in an amphitheater where the number of students ranged from 50 to 80 during the study. Indoor ambient air temperature, indoor relative humidity, and indoor air pressure were registered as well.

Air sampling system consisted of diaphragm vacuum pump Pfeiffer MVP, manifolds with openings for the continuous monitoring of inorganic gaseous pollutants (O_3, CO, SO_2, and NO_x) using Horiba devices (APOA, APMA, APSA, and APNA, 370 series) with 2-min resolution, and electronically controlled valves, which operated in alternating indoor/outdoor air sampling mode in the 10-min cycles. Over 24 h period, $PM_{2.5}$ was collected daily on quartz filters (Whatman QMA, 47 mm) using Svan Leckel LVS6-RV devices operating at a flow rate of 2.3 m^3 h^{-1}. The concentrations of $PM_{2.5}$ and their constituents, including trace elements (As, Cd, Cr, Mn, Ni, and Pb), ions (Cl^-, Na^+, Mg^{2+}, Ca^{2+}, K^+, NO_3^-, SO_4^{2-}, and NH_4^+) and 16 US EPA PAHs were determined at the reference laboratory of the Institute of Public Health of Belgrade. More details on the study area, sampling campaign, chemical analyses, quality assurance/quality control, and the relevant references are described previously (Stanišić et al., 2021).

Air Pollutant Sampling and Chemical Analyses

Inorganic gases (O_3, CO, SO_2, and NO_x) were continuously measured at indoor and outdoor sampling sites during a three-month campaign using Horiba devices (APOA, APMA, APSA, and APNA, 370 series) with 2-min resolution. The measurements based on ultraviolet absorption, infrared spectroscopy, ultraviolet fluorescence, and chemiluminescence methods for O_3, CO, SO_2, and NO_x, respectively, were performed according to the following Standards EN 14211:2012, EN 14212:2012, EN 14625:2012, and EN 14626:2012. The limit of detection (LOD) was 0.1 mg m^{-3} for CO and 1 μg m^{-3} for O_3, SO_2, and NO_x.

The radon concentrations (Bq m^{-3}) were measured by SN1029 radon monitor (Sun Nuclear Corporation, NRSB approval code 31822) with a time resolution of 30 min. The monitor contains two diffused junction photodiodes, which simultaneously detect radon, temperature, barometric pressure, and relative humidity. The LOD for radon was 0.1 Bq m^{-3}.

$PM_{2.5}$ was collected on quartz filters (Whatman QMA, 47 mm) for 24 h every day during a three-month campaign. For indoor and outdoor sampling, Svan Leckel LVS6-RV devices with a flow rate of 2.3 m^3 h^{-1} were used. Gravimetric measurements of $PM_{2.5}$ were performed according to the Standard EN 12341:2014. After preconditioning for 48 hours, the filters were weighed twice using a micro-balance (Precisa XR 125 SB) in Class 100 cleanroom and average values of $PM_{2.5}$ mass concentrations were further used. Prior to chemical analysis, filters from the sampling sites were stored in a cool room at 4°C. After gravimetric measurements, each filter (approximately 13.85 cm^2) was divided. Each half has a surface of approximately 1.76 cm^2 and one was used for the analysis of anions and cations (Cl^-, Na^+, Mg^{2+}, Ca^{2+}, K^+, NO_3^-, SO_4^{2-}, and NH_4^+). The remaining half 12.09 cm^2 were cut into two pieces which were used for the determination of trace elements (As, Cd, Cr, Mn, Ni, and Pb) and 16 US EPA PAHs.

For the determination of anion and cation concentrations, the filter extraction by ultra-pure water was carried out for 24 h and further analyzed by ion chromatography (Dionex DX500 IC system, MDL 064 Standard operating procedure). The LOD was as follows: 2 µg m^{-3} for Cl$^-$ and NO$_3^-$, 1 µg m^{-3} for SO$_4^{2-}$, 0.2 µg m^{-3} for NH$_4^+$, and 8 µg m^{-3} for Ca^{2+}.

The trace elements were determined following the Standard EN 14902:2005. A mixture of HNO$_3$ (30%):H$_2$O$_2$:H$_2$O (3:2:5) (analytical grade reagents, Merck) and distilled/deionized water (MiliQ, 18.2 MΩ) were used for the extraction (CEN/TC 264 N779) in a microwave accelerated digestor (Anton Paar 3000). The concentrations of As, Cd, Cr, Mn, Ni, and Pb were determined by inductively coupled plasma–mass spectrometry, ICP-MS (Agilent 7500ce with Octopole Reaction System). Standard reference material 2783 NIST (National Institute of Standard and Technology, MD, USA) was used for quality control and the recovery values were within a satisfactory range of ±20%. The LOD was as follows: 0.4 ng m^{-3} for As, 0.05 ng m^{-3} for Cd, and 2 ng m^{-3} for Cr, Mn, Ni, and Pb.

The concentrations of the following 16 priority PAHs (US EPA, 2005): naphthalene (Nap), acenaphthylene (Ace), acenaphthene (Ane), fluorene (Flu), phenanthrene (Phe), anthracene (Ant), fluoranthene (Fla), pyrene (Pyr), benzo[a]anthracene (B[a]A), chrysene (Chy), benzo[b]fluoranthene (B[b]F), benzo[k]fluoranthene (B[k]F), benzo[a]pyrene (B[a]P), dibenz[a,h]anthracene (DB[ah]A), benzo[g,h,i]perylene (B[ghi]P), and indeno[1,2,3-cd]pyrene (I[cd]P) were determined (ISO 12884:2010). Further details have previously been described (Stanišić et al., 2021; Cvetković et al., 2015). A mixture of n-hexane and acetone, 12.5 mL:12.5 mL (US EPA, 1999) was used for the microwave extraction of filters. Extracts were rotary evaporated to 1 mL under reduced pressure (55.6 kPa and with 0.2 ml isooctane) and to 0.25 mL under a nitrogen stream. Gas chromatography–mass selective detector (Agilent GC 6890/5973 MSD) with a DB-5 MS capillary column (30 m × 0.25 mm × 25 µm) was used for the analysis of PAHs (EPA Compendium Method TO-13A). The oven temperature was programmed as follows: (1) isothermal heating for 4 minutes at 70°C, (2) gradient heating from 70°C to 310°C at 8°C min^{-1}, and (3) isothermal heating for 5 minutes at 310°C. The solvent delay was 5 min and the time of run was 46 min. Helium was used as the carrier gas while the temperature of the injector was set to 300°C.

Prior to the analysis, calibration curves ($R^2 > 0.995$) with the concentration range between 5 and 200 ng mL^{-1} were obtained using Ultra Scientific PAH Mixture PM-831, which contains 16 US EPA PAHs. Ultra Scientific PAH Mixture PM-831, which consists of 16 compounds, each of 500.8 ± 2.5 µg mL^{-1} concentration was used as an external standard for the calibration curve. We used Ultra Scientific Semi-Volatiles Internal Standard Mixture ISM-560 (Ane-d10, Chy-d10, 1,4-dichlorobenzene, Nap-d8,

Perylene-d12, and Phe-d10) as an internal standard for the estimation of method recovery. Recovery values ranged from 85% to 110% for all the PAHs in the internal standard. The LOD was calculated as three times signal/noise and it was 0.01 ng m^{-3} while the limit of quantification was determined as 3.3 times of LOD. All data were corrected with reference to the field and laboratory blanks, which were prepared and analyzed as well.

Data Analysis

Machine Learning

To estimate the relationships between B[b]F levels and other PAHs, inorganic gaseous pollutants, trace elements, ions, and meteorological parameters in indoor and outdoor environments, the regression analysis was implemented using eXtreme Gradient Boosting. XGBoost is a technique of building a complex prediction model by iteratively combining ensembles of weak prediction models into a single strong learner possessing advantages such as computational efficiency and competitive accuracy, even when sparse and unstructured data are used (Hartmann, 2019, Lundberg et al., 2020). It builds a sequential series of smaller decision trees, where each tree efforts to complement all others and correct for the residuals in the predictions made by all previous trees (Sheridan et al., 2016). In this study, we used Python (Python Software Foundation) XGBoost implementation (XGBoost Python Package). The dataset was split into training (80%) and validation (20%) sets. Hyperparameter tuning was implemented using a brute-force grid search and stratified cross-validation that was replicated ten times. The best performing hyperparameter values were used for the final model.

Explainable Artificial Intelligence

The explainability of the ML model behavior that operates with high-dimensional input data in a non-linear and nested fashion is crucial for understanding the process being modeled. For this purpose, we employed the advanced explainable artificial intelligence method, which is capable to provide a straightforward and meaningful interpretation of ML model-derived decisions, now being shifted towards user-readable logic rules (Stanišić et al., 2021).

SHapley Additive exPlanations

SHapley Additive exPlanations is a method based on Shapley values, calculated as a measure of feature importance using a game-theory approach (Lundberg and Lee, 2017). In this study we used Python SHAP implementation (SHAP Python package). The captured attributed importance of a feature, the change of feature importance over its value range, and its interaction effects with other features are visually presented as SHAP summary plots, SHAP dependency plots, and SHAP interaction plots, respectively.

Fuzzy Clustering

The fuzzy clustering of absolute SHAP attributions was performed to identify and characterize indoor and outdoor ambient conditions responsible for B[b]F behavior. It was chosen because each event will not necessarily belong to a single class of environmental conditions that shape it. Fuzzy clustering was performed by using an R 'cluster' package (Maechler et al., 2019). The obtained results were presented as force plots. A detailed analysis of each cluster was performed based on the statistical character of its absolute and relative SHAP values, as well as the measured parameter values.

Results and Discussion

Descriptive Statistics

The results have shown that the mean B[b]F level was negligibly higher in the outdoor environment (0.9 ng m^{-3}) compared to the indoor (0.7 ng m^{-3}). Thereby, minimum B[b]F levels were similar both indoor and outdoor (0.08 vs. 0.10 ng m^{-3}), while the maximum levels were registered indoor (7.1 vs. 3.8 ng m^{-3}). The study of Živković et al., (2015) made a comparison of total gas and particle-bound B[b]F concentrations registered at different school locations throughout the year. The average winter levels were in the range from 1.72 to 9.11 ng m^{-3}, and from 3.36 to 13.68 ng m^{-3}, for indoor and outdoor environments, respectively, while the concentrations were significantly lower during the warm season, ranging from 0.15 to 1.49 ng m^{-3}.

According to the Pearson's correlation analysis, indoor B[b]F levels significantly correlated (r>0.90) with the levels of B[a]A, Chy, I[cd]P, B[ghi]P, B[a]P, B[k]F and Pyr, while in the outdoor environment B[b]F levels exhibited significant correlations (r > 0.90) with B[a]A, Chy, I[cd]P, B[a]P, B[k]F, and B[ghi]P levels. While the listed pollutants were observed to share common sources and display similar behavior patterns within the environment they were analyzed in, the correlation between indoor and outdoor B[b]F levels was not too high (0.62; p < 0.05). The outdoor PAH air burden can be associated with emissions from fossil fuel combustion for numerous purposes, including transport, and have an impact on indoor air quality. Nevertheless, in addition to this, indoor PAH levels are influenced by the reemission, resuspension, and sorption of PAHs from indoor dust and human activities (Morawska et al., 2013; Stanišić et al., 2021). Furthermore, no significant correlation was observed between B[b]F levels and meteorological variables, concentrations of inorganic gases, trace elements, and ions. In addition to this, the number of people, the time they spent indoors, temporal trends in concentrations, weekday, and weekend were also found to be irrelevant for predicting B[b]F levels.

In order to further investigate the relationships between B[b]F and other variables, we have amended these findings with the results of the advanced machine learning methods. For this purpose, XGBoost was successfully employed for the exploration of non-linear relationships between B[b]F levels and key variables that shape its distribution in indoor and outdoor air. As indicated by the predicted and observed relative errors (12.5% and 10.5%, respectively) and high correlation coefficients (r^2 = 0.97 and 0.98 for indoor and outdoor, respectively), the model was successfully applied (Figure 1).

As shown by SHAP analysis, the most important variables that shaped B[b]F levels in the indoor environment were attributed the highest positive (up to 0.8) and negative (up to -0.2) SHAP values (Figures 2 and 3), according to the following order: B[k]F, Chy, B[a]A, I[cd]P, B[a]P, Pyr, Rh (relative humidity), CO, and Fla. Besides, the analysis revealed less significant impacts of D[ah]A, B[ghi]P, and inorganic elemental and ionic $PM_{2.5}$ constituent levels (Cr, Rn, As, SO_4^{2-}, and NH_4^+) on B[b]F behavior patterns in indoor environment. The similar applies to the outdoor environment as the highest impacts on the B[b]F distribution, described by SHAP values (from -0.15 to 0.5) were attributed to the following compounds: B[a]P, B[a]A, B[k]F, Chy, I[cd]P, and B[ghi]P. Less significant effects on B[b]F outdoor levels (SHAP = -0.1 – 0.1) were attributed to both PAHs and inorganic pollutants (Pyr, CO, NH_4^+, Phen, As, NO_3^-, Fla, Mn, D[ah]A, Cr, and Ant), as well as to meteorological parameters including cinh (convective inhibition), WD (wind direction), temperature (temp), and lib4 (best 4-layer lifted index).

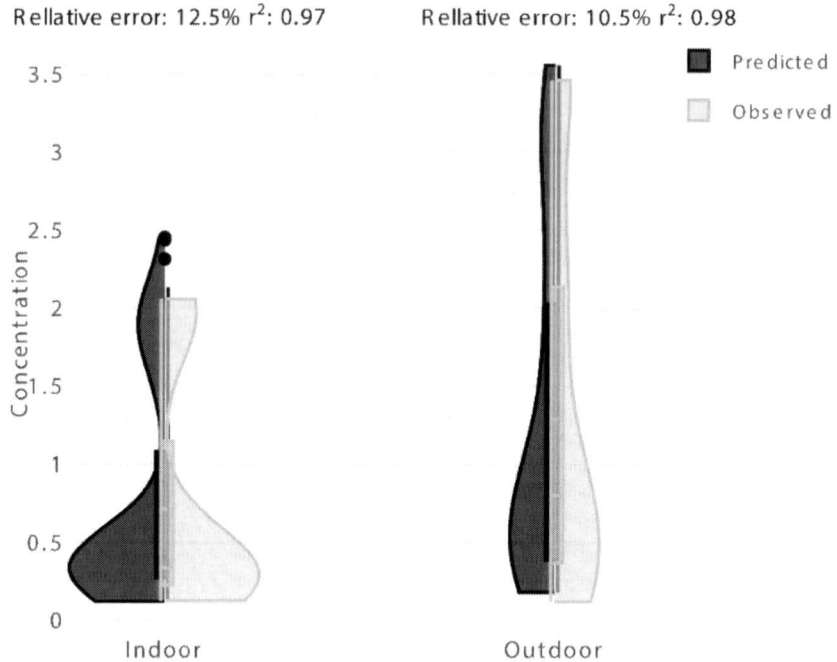

Figure 1. XGBoost model evaluation.

Coal combustion and related pyrogenic processes were identified as dominant sources that contributed to the registered levels of B[b]F, B[k]F, Chy, B[a]A, B[a]P, I[cd]P, and B[ghi]P at the measurement site. As indicated by SHAP dependence plots, the highest impacts of CO, As, Cr, and Mn, corresponding to the highest concentrations of these compounds, suggest common sources and processes which shape B[b]F levels.

The increases in the concentrations of ionic species (outdoor NH_4^+: up to 11.9 μg m^{-3}; indoor NH_4^+: up to 7.2 μg m^{-3}; outdoor NO_3^-: up to 17 μg m^{-3}; and indoor SO_4^{2-}: up to 6.9 μg m^{-3};) had a positive impact on B[b]F levels both indoor and outdoor. As indicated by SHAP dependence plots (Figure 4 and 5), higher concentrations of nitrate ions (> 8 μg m^{-3}) were related to lower B[b]F levels (< 2 ng m^{-3}), whereas the increase in SO_4^{2-} and NH_4^+ levels corresponded to the increase in B[b]F concentrations.

As regards meteorological parameters, the impact of northern winds, convective inhibition and lifted index on B[b]F outdoor levels was evidenced (cinh: up to 0 J kg^{-1}; lib4: up to 10.5 K), as well as the minor impact of lower temperatures (down to 4.8°C). The impact of wind direction on PM$_{2.5}$ concentrations and the associated compounds depends on local geographical and topographic conditions and may lead to the reduction or enhanced accumulation of pollutants (Chen et al., 2020). The relationship between wind direction and outdoor B[b]F levels confirmed the findings of previous studies that have shown that, in addition to many individual local emissions, long-range transport, and emissions from distant sources including thermal power plants, petrochemical industry, and oil refineries affect air quality over the study area (Perišić et al., 2017; Stojić et al., 2016). The relationship between B[b]F concentrations and convective inhibition, which represents a measure of the energy required by the atmosphere to inhibit the rising of air, can be explained by the fact that vertical air mass movements affect the retention of particle-bound PAHs in the ground layers. Besides, under low-temperature conditions, atmospheric convection weakens and enhances the accumulation of PM$_{2.5}$. In contrast to this, high temperatures would induce turbulent movements that would further accelerate the dispersion of PM$_{2.5}$ (Li et al., 2015; Yang et al., 2016). Finally, the importance of the positive lifted index suggests that the outdoor B[b]F concentration dynamics of is shaped by stable tropospheric conditions with respect to boundary layer-based convection.

In the indoor environment, a variable impact of relative humidity on B[b]F levels has been shown. Namely, a negative impact on B[b]F levels corresponded with higher values of relative humidity (up to 57%), probably because the gas to particle partitioning of pollutants is favored under these conditions, as well as the increase in PM$_{2.5}$ concentrations and the uptake of B[b]F and other high-ring PAHs onto particles with high organic content (Wang and Ogawa, 2015; Liao et al., 2017).

In contrast to this, lower humidity (down to 20%) had a positive impact on B[b]F levels because it led to an enhanced evaporation loss of PM$_{2.5}$ and associated contaminants (Liu et al., 2015). Madruga et al., (2019) also emphasized relative humidity as a prominent factor, which has an impact on the PAH behavior.

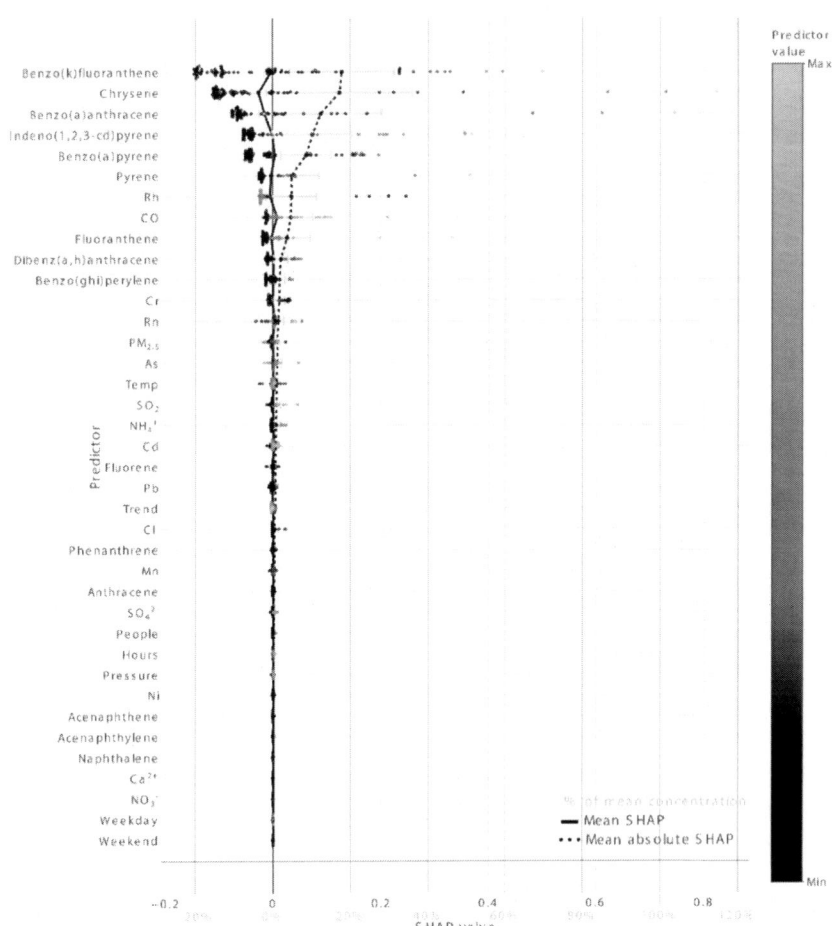

Figure 2. Indoor B[b]F SHAP summary plot.

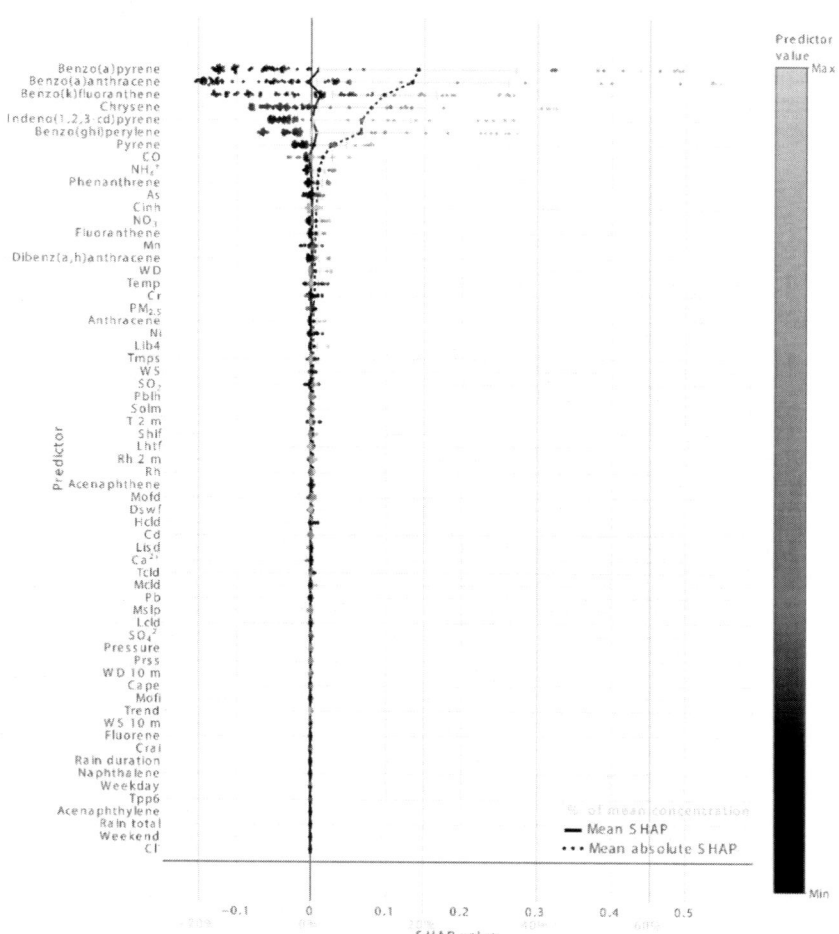

Figure 3. Outdoor B[b]F SHAP summary plot.

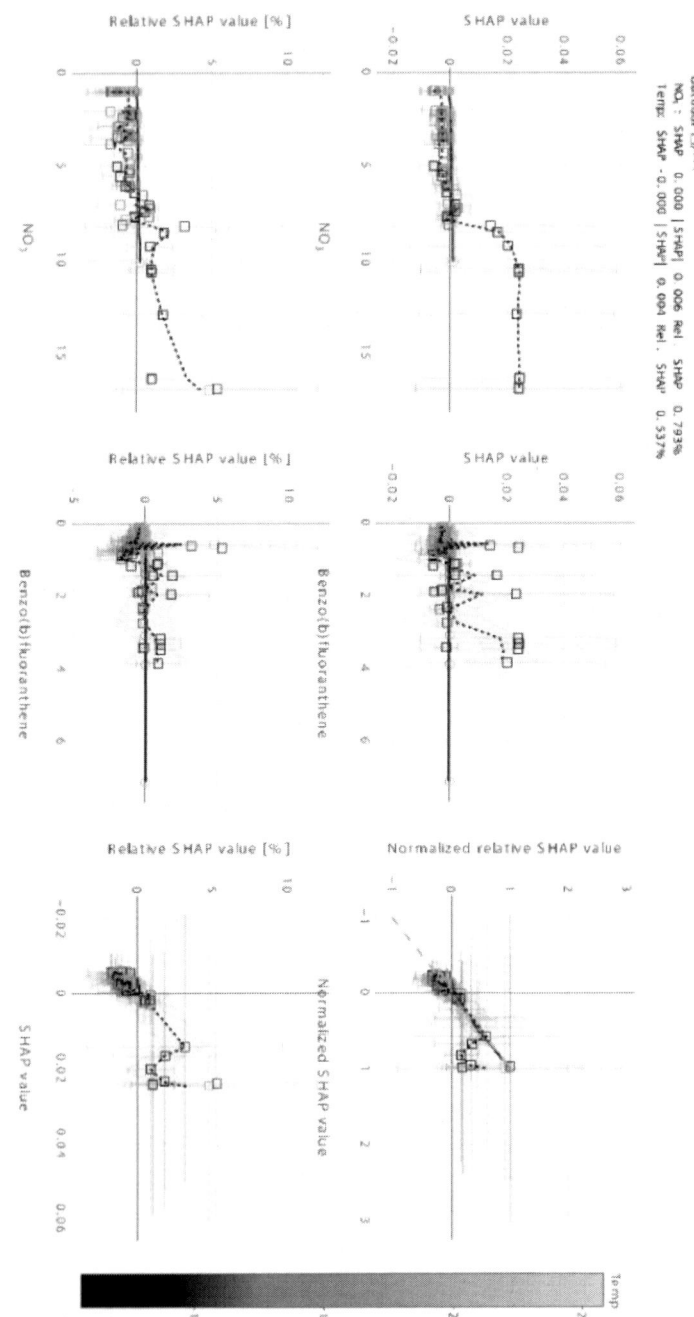

Figure 4. B[b]F SHAP dependence on NO₃⁻ and temperature.

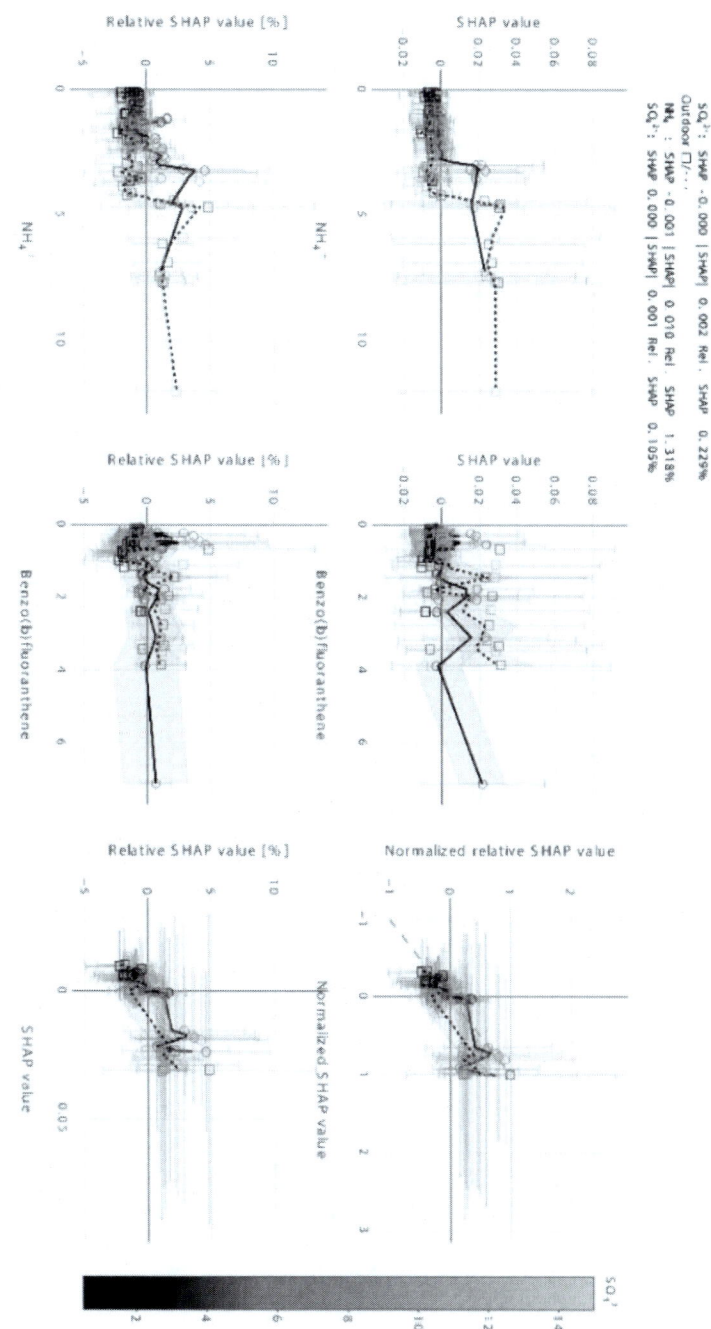

Figure 5. B[b]F SHAP dependence on NH_4^+ and SO_4^{2-}.

Interactions between Environmental Factors and Indoor and Outdoor B[b]F

Although many investigations evidenced that indoor air quality depends on outdoor pollutant concentrations, it does not represent a simple reflection of the outdoor ambiance. Figure 6 represents the fuzzy clustering of SHAP values, according to which eight groups of variables that shape B[b]F dynamics in indoor and outdoor (relative error < 25% and 30%, respectively) environments were identified.

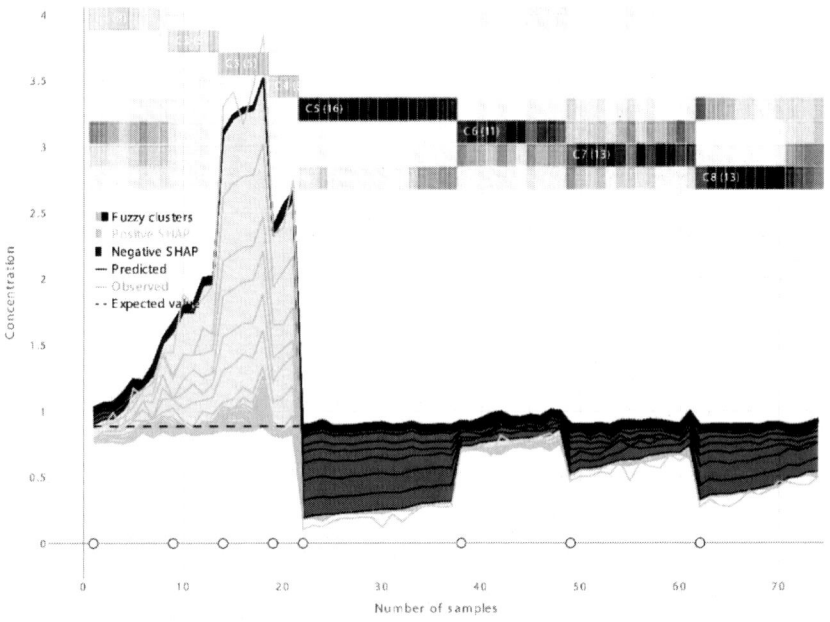

Figure 6. Outdoor B[b]F SHAP force plot.

The constituents of four clusters (°C1: March 4th – April 30th; °C2: March 7th – April 11th; °C3: March 10th – March 25th; and °C4: March 20th – May 17th) had a positive impact on outdoor B[b]F levels from the mean to 95th quantile values (0.87–3.31 ng m^{-3}). Thereby, the highest impacts indicated by relative SHAP values (%) were attributed to the following PAHs: B[a]A, B[k]F, Chy, B[a]P, I[cd]P, B[ghi]P, and Pyr. Additionally, minor impacts were revealed for convective inhibition, NH_4^+, NO_3^-, and As. The highest B[b]F concentrations during the period assigned to °C3 were accompanied by the increased levels of the pollutants (B[ghi]P: 3.44 ng m^{-3}; Chy: 2.73 ng m^{-3}; B[k]F: 2.33 ng m^{-3}; B[a]A: 2.04 ng m^{-3}; B[a]P: 2.02 ng m^{-3}; I[cd]P: 2.00 ng m^{-3}; Pyr: 0.95 ng m^{-3}; As: 1.84 ng m^{-3}; NH$_4^+$: 6.18 µg m^{-3}; and NO$_3^-$: 10.66 µg m^{-3}) originating from the joint pollution sources. Low-temperature periods were associated with intensified pollutant emissions from heating sources. Furthermore, the absence of PAH photolytic degradation also contributed to the elevated outdoor pollutant levels (Stanišić et al., 2021). The periods assigned to ºC1, ºC2, and ºC4 were characterized by changeable weather conditions with occasional precipitations, strong winds, and penetration of warmer and dry air masses. During these periods, reduced activities of heating plants and individual fireboxes using coal and wood, as well as dispersion and photooxidative degradation caused by higher temperatures and stronger winds led to a decrease in PAH outdoor levels (from 0.4 ng m^{-3} to 2.5 ng m^{-3}, belonging from 25th to 75th quantile values).

Fuzzy clustering also identified clusters (ºC5: March 26th – May 31st; ºC6: March 30th – May 20th; ºC7: April 1st – May 24th; and ºC8: April 7th – May 25th) representing the groups of variables related to 25th to 75th quantile outdoor B[b]F values (from 0.27 ng m^{-3} to 0.28 ng m^{-3}) (Figure 6). The same predictors as discussed above shaped the clusters, but relative SHAP values (%) revealed negative interrelations between B[b]F and certain pollutant (B[a]A, B[k]F, Chy, B[a]P, I[cd]P, B[ghi]P, and CO) behavior patterns. The presence of CO can be attributed to burning wood or coal in individual heating units in densely populated residential areas surrounding the measurement site. The contribution of traffic emissions to PAH burden was lower compared to the contributions of fossil fuel burning for heating operations, and it appeared to remain stable over the measurement campaign.

Four clusters (^{i}C1: March 4th – May 9th; ^{i}C4: March 7th – March 24th; ^{i}C5: March 10th – March 28th; and ^{i}C6: March 18th – March 20th), grouping variables with a positive impact on B[b]F indoor distribution, were identified (Figure 7). Two of them (^{i}C5 and ^{i}C6) were more differentiated with the highest B[b]F indoor levels (from 1.41 to 7.11 ng m^{-1}). The periods assigned to these clusters corresponded to ºC3, which implies the association between outdoor and indoor air quality. During the mentioned periods, the episodes of cold, dry weather (temperature below 10°C) were linked to the intensified emissions from heating sources. As shown by relative SHAP values (%), the higher indoor B[b]F concentrations were shaped by the increased levels of 4- (Chy and B[a]A; Fla and Pyr), 5 (B[a]P and B[k]F), and 6-ring (B[ghi]P and I[cd]P) structural PAH isomers. The impact of fossil fuel combustion for heating purposes was additionally confirmed by non-negligible attributions of

CO to certain clusters (^{i}C4: 4%; ^{i}C5: 6%; and ^{i}C6: 6%) related to the elevated concentration of this gas (^{i}C4: 0.82 mg m^{-3}; ^{i}C5: 1.13 mg m^{-3}; and ^{i}C6: 5.53 mg m^{-3}). The impact of relative humidity increased with the increase of its value. In general, relative humidity was twice lower indoor (36.8%) than outdoor (61.5%) during the measurement campaign, due to the heating of indoor spaces. Under reduced humidity conditions, evaporation of pollutants from particle surface may occur (Liu et al., 2015) in contrast to their enhanced adsorption/absorption in circumstances of higher relative humidity. This was evident in a significantly lower abundance of semi-volatile Chy in ^{i}C4 compared to ^{i}C6, later being characterized by higher humidity.

Low B[b]F indoor concentrations (< 0.53 ng m^{-3}) indoor were associated with four clusters (^{i}C2: March 5th – May 26th; ^{i}C3: March 6th – May 24th; ^{i}C7: April 10th – May 27th; and ^{i}C8: April 15th – May 23rd) which comprised the negative impacts of B[a]A, Chy, B[a]P, Fla, I[cd]P, Pyr, B[k]F, D[ah]A, relative humidity and Rn on B[b]F indoor behavior. The strength of their effects is indicated by the relative SHAP values (%). It is worth noting that the four analyzed clusters were related to the lower occurrence of B[a]A, Chy, B[a]P, Fla, I[cd]P, Pyr, B[k]F, D[ah]A (< 0.5 ng m^{-3}) compared to ^{i}C1, ^{i}C4, ^{i}C5, and ^{i}C6.

Radon is usually presented indoors as a buildup impurity originating from underlying soil, building material and its granular size, pore air spaces, and moisture content (Burghele et al., 2021). In a microstructure level, Rn occupies pore spaces inside mineral surface layers and could migrate by diffusion and advection until it releases from the opened pathways into the atmosphere (Syuryavin et al., 2020). The low positive impact of Rn, occurring in high concentrations (> 70 Bq m^{-3}), implies that dust resuspension, containing both particle-bound and pore-entrapped pollutants, is a non-negligible pollution source in the indoor environment apart from emissions of fossil fuel combustion for the industrial, heating and traffic purposes, and pollutant diffusion from outdoors.

Generally, the clusters distinguished by high to excessive concentrations of PAHs could be considered the most influential for the observed high outdoor and indoor B[b]F levels, as indicated by the absolute and relative SHAP values.

Absolute and relative SHAP interaction values implied the most significant interactions among the studied variables that shape indoor and outdoor B[b]F behaviour. The most important pairs of predictors indoors were: B[a]P–I[cd]P, B[k]F–SO$_4^{2-}$, Chy–Cr, I[cd]P–CO, D[ah]P–NH$_4^+$, I[cd]P–Pb, Cr–Pb, B[a]A–CO, B[k]F–CO, Chy–CO, Chy–B[a]P, Chy–B[k]F,

Chy–PM$_{2.5}$, Chy–Rn, and Rh–CO, while in the outdoor environment the following pairs were recognized as important: B[a]P–B[a]A, B[a]P–Chy, B[a]P–B[k]F, B[k]F–B[a]A, I[cd]P–B[a]A, I[cd]P–B[a]P, I[cd]P–B[ghi]P, I[cd]P–Chy, B[ghi]P–B[a]A, and B[ghi]P–CO.

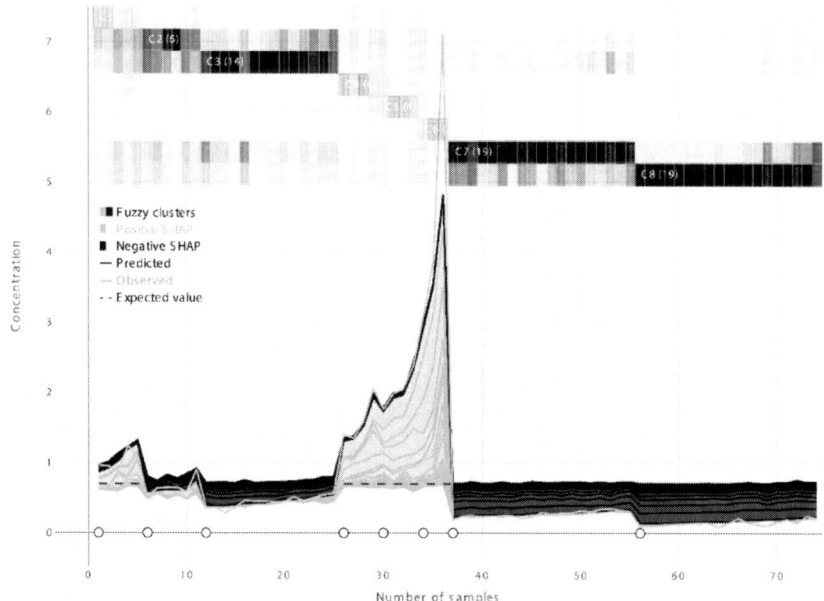

Figure 7. Indoor B[b]F SHAP force plot.

Acknowledgments

Funding: The authors acknowledge funding provided by the Institute of Physics Belgrade, through the grant by the Ministry of Education, Science and Technological Development of the Republic of Serbia, the Science Fund of the Republic of Serbia #GRANT No. 6524105, AI – ATLAS.

Conclusion

Although B[a]P is commonly used as a marker of exposure to mixtures of carcinogenic PAHs in indoor and outdoor environments, emerging research points the need to shift towards the investigation of atmospheric behavior of other health-relevant possible/probable carcinogenic PAHs. This chapter resumes the finding based on the application of machine learning and explainable artificial intelligence (as indicated by the model evaluation statistics) for investigating the relationships between B[b]F levels and dozens of variables including inorganic and organic $PM_{2.5}$ constituents, meteorological parameters, and the number of people in the amphitheater, and the time they spent indoor. According to the results, high molecular weight particle-bound PAHs (B[a]P, B[k]F, B[ghi]P, I[cd]P, and D[ah]A), followed by 4-ring isomers (Chy and B[a]A; Fla and Pyr) were dominantly related with the presence of B[b]F in both indoor and outdoor environments. Under stable tropospheric conditions, temperature, convective inhibition, wind direction, and lifted index exhibited less significant impacts on B[b]F fate outdoors while relative humidity appeared to be the most distinctive meteorological parameter indoors. Less important functional dependencies were observed between B[b]F and inorganic pollutants (CO, Rn, As, Cr, SO_4^{2-}, NH_4^+, and NO_3^-) suggesting their similar origin and participation in oxidative degradation and aerosol formation. The herein presented explainable artificial intelligence methodology represents a promising tool for exploring and understanding the complexity of environmental conditions which shape air pollution processes.

References

Bansal, V. and Kim, K. H. (2015). Review of PAH contamination in food products and their health hazards. *Environment International*, 84, 26-38.

Blair, G. S., Henrys, P., Leeson, A., Watkins, J., Eastoe, E., Jarvis, S. and Young, P. J. (2019). Data science of the natural environment: a research roadmap. *Frontiers in Environmental Science*, 7, 121-129.

Brown, A. S., Sarantaridis, D., Butterfield, D. M., Brown, R. J., Whiteside, K. J., Hughey, P., Goddard, S. L., Hussain, D. and Williams, M. (2012). *Annual Report for 2011 on the UK PAH Monitoring and Analysis Network*.

Burghele, B. D., Botoș, M., Beldean-Galea, S., Cucoș, A., Catalina, T., Dicu, T., Dobrei, G., Florică, Ș., Istrate, A., Lupulescu, A. and Moldovan, M. (2021). Comprehensive survey on radon mitigation and indoor air quality in energy efficient buildings from Romania. *Science of The Total Environment*, 751, 141858.

Cao, Z., Wang, M., Chen, Q., Zhu, C., Jie, J., Li, X., Dong, X., Miao, Z., Shen, M. and Bu, Q. (2019). Spatial, seasonal and particle size dependent variations of PAH contamination in indoor dust and the corresponding human health risk. *Science of the Total Environment*, 653, 423-430.

Chen, H., Lundberg, S. and Lee, S. I. (2021). Explaining models by propagating Shapley values of local components. In *Explainable AI in Healthcare and Medicine*, 261-270. Springer, Cham.

Chen, Z., Chen, D., Zhao, C., Kwan, M. P., Cai, J., Zhuang, Y., Zhao, B., Wang, X., Chen, B., Yang, J. and Li, R. (2020). Influence of meteorological conditions on $PM_{2.5}$ concentrations across China: A review of methodology and mechanism. *Environment International*, 139, 105558.

Cvetković, A., Jovašević-Stojanović, M., Marković, D. and Ristovski, Z. (2015). Concentration and source identification of polycyclic aromatic hydrocarbons in the metropolitan area of Belgrade, Serbia. *Atmospheric Environment*, 112, 335-343.

Elorduy, I., Elcoroaristizabal, S., Durana, N., García, J. A. and Alonso, L. (2016). Diurnal variation of particle-bound PAHs in an urban area of Spain using TD-GC/MS: influence of meteorological parameters and emission sources. *Atmospheric Environment*, 138, 87-98.

European Standards (EN) 12341:2014 (2014). *Ambient air. Standard gravimetric measurement method for the determination of the PM_{10} or $PM_{2.5}$ mass concentration of suspended particulate matter.* https://cds.cern.ch/record/ 2624772 (Last accessed February 4, 2021).

European Standards (EN) 14211:2012 (2012). *Ambient air. Standard method for the measurement of the concentration of nitrogen dioxide and nitrogen monoxide by chemiluminescence.* https://www.en-standard.eu/bs-en-14211-2012-ambient-air.-standard-method-for-the-measurement-of-the-concentration-of-nitrogen-dioxide-and-nitrogen-monoxide-by-chemiluminescence/ (Last accessed February 4, 2021).

European Standards (EN) 14212:2012 (2012). *Ambient air. Standard method for the measurement of the concentration of sulphur dioxide by ultraviolet fluorescence.* https://www.sis.se/en/produkter/ environment-health-protection-safety/air-quality/ambient-atmospheres/ssen142122012/ (Last accessed February 4, 2021).

European Standards (EN) 14625:2012 (2012). *Ambient air. Standard method for the measurement of the concentration of ozone by ultraviolet photometry.* https://shop.bsigroup.com/ProductDetail? pid=000000000030210754 (Last accessed February 4, 2021).

European Standards (EN) 14626:2012 (2012). *Ambient air. Standard method for the measurement of the concentration of carbon monoxide by non-dispersive infrared spectroscopy.* https://www.en-standard.eu/bs-en-14626-2012-ambient-air.-standard-method-for-the-measurement-of-the-concentration-of-carbon-monoxide-by-non-dispersive-infrared-spectroscopy/ (Last accessed February 4, 2021).

European Standards (EN) 14902:2005 (2005). *Ambient air quality. Standard method for the measurement of Pb, Cd, As, and Ni in the PM_{10} fraction of suspended particulate matter.* https://www. antpedia.com/standard/pdf/ 13.040.20/1703/EN%2014902-2005_ 6234.pdf (Last accessed February 4, 2021).

Ferguson, L., Taylor, J., Davies, M., Shrubsole, C., Symonds, P. and Dimitroulopoulou, S. (2020). Exposure to indoor air pollution across socio-economic groups in high-income countries: A scoping review of the literature and a modelling methodology. *Environment International*, 143, 105748.

Friesen, M. C., Demers, P. A., Spinelli, J. J., Eisen, E. A., Lorenzi, M. F. and Le, N. D. (2010). Chronic and acute effects of coal tar pitch exposure and cardiopulmonary mortality among aluminum smelter workers. *American Journal of Epidemiology*, 172, 790-799.

Gaga, E. O. and Ari, A. (2019). Gas-particle partitioning and health risk estimation of polycyclic aromatic hydrocarbons (PAHs) at urban, suburban and tunnel atmospheres: Use of measured EC and OC in model calculations. *Atmospheric Pollution Research*, 10, 1-11.

Ghosal, D., Ghosh, S., Dutta, T. K. and Ahn, Y. (2016). Current state of knowledge in microbial degradation of polycyclic aromatic hydrocarbons (PAHs): a review. *Frontiers in Microbiology*, 7, 1369-1382.

Gitipour, S., Sorial, G. A., Ghasemi, S. and Bazyari, M. (2018). Treatment technologies for PAH-contaminated sites: a critical review. *Environmental Monitoring and Assessment*, 190, 1-17.

Gohlke, J. M., Doke, D., Tipre, M., Leader, M. and Fitzgerald, T. (2011). A review of seafood safety after the Deepwater Horizon blowout. *Environmental Health Perspectives*, 119, 1062-1069.

González-Martín, J., Kraakman, N., Pérez, C., Lebrero, R. and Muñoz, R. (2020). A state-of-the-art review on indoor air pollution and strategies for indoor air pollution control. *Chemosphere*, 262, 128376.

Gune, M. M., Ma, W. L., Sampath, S., Li, W., Li, Y. F., Udayashankar, H. N., Balakrishna, K. and Zhang, Z. (2019). Occurrence of polycyclic aromatic hydrocarbons (PAHs) in air and soil surrounding a coal-fired thermal power plant in the south-west coast of India. *Environmental Science and Pollution Research*, 26, 22772-22782.

Hao, X., Li, J. and Yao, Z. (2016). Changes in PAHs levels in edible oils during deep-frying process. *Food Control*, 66, 233-240.

Haritash, A. K. and Kaushik, C. P. (2009). Biodegradation aspects of polycyclic aromatic hydrocarbons (PAHs): a review. *Journal of Hazardous Materials*, 169, 1-15.

Hitzel, A., Pöhlmann, M., Schwägele, F., Speer, K. and Jira, W. (2013). Polycyclic aromatic hydrocarbons (PAH) and phenolic substances in meat products smoked with different types of wood and smoking spices. *Food Chemistry*, 139, 955-962.

Hussar, E., Richards, S., Lin, Z. Q., Dixon, R. P. and Johnson, K. A. (2012). Human health risk assessment of 16 priority polycyclic aromatic hydrocarbons in soils of Chattanooga, Tennessee, USA. *Water, Air, & Soil Pollution*, 223, 5535-5548.

International Agency for Research on Cancer (2012). A review of human carcinogens. Part F: Chemical agents and related occupations. *IARC monographs on the evaluation of carcinogenic risks to humans.* https://www. thelancet.com/journals/lanonc/article/ PIIS1470-2045(09)70358-4/fulltext (Last accessed: February 2, 2021).

International Standardization Organization (ISO) 12884:2000 (2020). Ambient air — Determination of total (gas and particle-phase) polycyclic aromatic hydrocarbons — *Collection on sorbent-backed filters with gas chromatographic/mass spectrometric analyses.* https://www.iso.org/ standard/ 1343.html (Last accessed February 4, 2021).

Johnsen, A. R., Wick, L. Y. and Harms, H. (2005). Principles of microbial PAH-degradation in soil. *Environmental Pollution*, 133, 71-84.

Jovanović, G., Romanić, S. H., Stojić, A., Klinčić, D., Sarić, M. M., Letinić, J. G. and Popović, A. (2019). Introducing of modeling techniques in the research of POPs in breast milk–A pilot study. *Ecotoxicology and environmental safety*, 172, 341-347.

Kim, K. H., Jahan, S. A., Kabir, E. and Brown, R. J. (2013). A review of airborne polycyclic aromatic hydrocarbons (PAHs) and their human health effects. *Environment International*, 60, 71-80.

Lee, J. G., Kim, S. Y., Moon, J. S., Kim, S. H., Kang, D. H. and Yoon, H. J. (2016). Effects of grilling procedures on levels of polycyclic aromatic hydrocarbons in grilled meats. *Food Chemistry*, 199, 632-638.

Lee, R. F. (2003). Photo-oxidation and photo-toxicity of crude and refined oils. *Spill Science & Technology Bulletin*, 8, 157-162.

Li, J., Chen, H., Li, Z., Wang, P., Cribb, M. and Fan, X. (2015). Low-level temperature inversions and their effect on aerosol condensation nuclei concentrations under different large-scale synoptic circulations. *Advances in Atmospheric Sciences*, 32, 898-908.

Liao, T., Wang, S., Ai, J., Gui, K., Duan, B., Zhao, Q., Zhang, X., Jiang, W. and Sun, Y. (2017). Heavy pollution episodes, transport pathways and potential sources of $PM_{2.5}$ during the winter of 2013 in Chengdu (China). *Science of the Total Environment*, 584, 1056-1065.

Liu, C. N., Lin, S. F., Tsai, C. J., Wu, Y. C. and Chen, C. F. (2015). Theoretical model for the evaporation loss of $PM_{2.5}$ during filter sampling. *Atmospheric Environment*, 109, 79-86.

Lowor, S. T., Jacquet, M., Vrielink, T., Aculey, P., Cros, E. and Takrama, J. (2012). Post-harvest sources of polycyclic aromatic hydrocarbon contamination of cocoa beans: a simulation. *International Journal of AgriScience*, 2, 1043-1052.

Lundberg, S. M., Erion, G., Chen, H., DeGrave, A., Prutkin, J. M., Nair, B., Katz, R., Himmelfarb, J., Bansal, N. and Lee, S. I. (2020). From local explanations to global understanding with explainable AI for trees. *Nature Machine Intelligence*, 2, 56-67.

Lundberg, S. M., Lee, S. I. (2017). A unified approach to interpreting model predictions, in: Guyon, I., Luxburg, U. V., Bengio, S., Wallach, H., Fergus, R., Vishwanathan, S., Garnett, R. (Eds.), *Advances in Neural Information Processing Systems 30* (NIPS 2017), 4765-4774.

Lundstedt, S., White, P. A., Lemieux, C. L., Lynes, K. D., Lambert, I. B., Öberg, L., Haglund, P. and Tysklind, M. (2007). Sources, fate, and toxic hazards of oxygenated polycyclic aromatic hydrocarbons (PAHs) at PAH-contaminated sites. *AMBIO: A Journal of the Human Environment*, 36, 475-485.

Madruga, D. G., Ubeda, R. M., Terroba, J. M., Dos Santos, S. G. and García-Cambero, J. P. (2019). Particle-associated polycyclic aromatic hydrocarbons in a representative urban location (indoor-outdoor) from South Europe: assessment of potential sources and cancer risk to humans. *Indoor Air*, 29, 817-827.

Maechler, M., Rousseeuw, P., Struyf, A., Hubert, M., Hornik, K. (2019). Cluster: Cluster Analysis Basics and Extensions. R package version 2.1.0. https://cran.r-project.org/web/packages/cluster/cluster.pdf (Last accessed: January 27, 2021).

Morawska, L., Afshari, A., Bae, G. N., Buonanno, G., Chao, C. Y. H., Hänninen, O., Hofmann, W., Isaxon, C., Jayaratne, E. R., Pasanen, P. and Salthammer, T. (2013). Indoor aerosols: from personal exposure to risk assessment. *Indoor Air*, 23, 462-487.

Oliveira, M., Slezakova, K., Delerue-Matos, C., Pereira, M. C. and Morais, S. (2019). Children environmental exposure to particulate matter and polycyclic aromatic hydrocarbons and biomonitoring in school environments: a review on indoor and outdoor exposure levels, major sources and health impacts. *Environment International*, 124, 180-204.

Parajuli, A., Grönroos, M., Kauppi, S., Płociniczak, T., Roslund, M. I., Galitskaya, P., Laitinen, O. H., Hyöty, H., Jumpponen, A., Strömmer, R. and Romantschuk, M. (2017). The abundance of health-associated bacteria is altered in PAH polluted soils—Implications for health in urban areas? *PLoS One*, 12, e0187852.

Paris, A., Ledauphin, J., Poinot, P. and Gaillard, J. L. (2018). Polycyclic aromatic hydrocarbons in fruits and vegetables: Origin, analysis, and occurrence. *Environmental Pollution*, 234, 96-106.

Paulik, L. B., Hobbie, K. A., Rohlman, D., Smith, B. W., Scott, R. P., Kincl, L., Haynes, E. N. and Anderson, K. A. (2018). Environmental and individual PAH exposures near rural natural gas extraction. *Environmental Pollution*, 241, 397-405.

Pereira, D. C. A., Custódio, D., de Andrade, M. D. F., Alves, C. and de Castro Vasconcellos, P. (2019). Air quality of an urban school in São Paulo city. *Environmental Monitoring and Assessment*, 191, 1-13.

Perišić, M., Rajšić, S., Šoštarić, A., Mijić, Z. and Stojić, A. (2017). Levels of PM_{10}-bound species in Belgrade, Serbia: spatio-temporal distributions and related human health risk estimation. *Air Quality, Atmosphere & Health*, 10, 93-103.

Rostami, R., Kalan, M. E., Ghaffari, H. R., Saranjam, B., Ward, K. D., Ghobadi, H., Poureshgh, Y. and Fazlzadeh, M. (2021). Characteristics and health risk assessment of heavy metals in indoor air of waterpipe cafés. *Building and Environment*, 190, 107557.

Rotkin-Ellman, M., Wong, K. K. and Solomon, G. M. (2012). Seafood contamination after the BP Gulf oil spill and risks to vulnerable populations: a critique of the FDA risk assessment. *Environmental Health Perspectives*, 120, 157-161.

Samburova, V., Zielinska, B. and Khlystov, A. (2017). Do 16 polycyclic aromatic hydrocarbons represent PAH air toxicity? *Toxics*, 5, p. 17.

Sievert, C. (2020). *Interactive Web-Based Data Visualization with R, plotly, and shiny*. CRC Press.

Slezakova, K., Pires, J. C. M., Castro, D., Alvim-Ferraz, M., Delerue-Matos, C., Morais, S. and Pereira, M. D. C. (2013). PAH air pollution at a Portuguese urban area: carcinogenic risks and sources identification. *Environmental Science and Pollution Research*, 20, 3932-3945.

Stanišić, S., Perišić, M., Jovanović, G., Milićević, T., Romanić, S. H., Jovanović, A., Šoštarić, A., Udovičić, V. and Stojić, A. (2021). The $PM_{2.5}$-bound polycyclic aromatic hydrocarbon behavior in indoor and outdoor environments, part I: Emission sources. *Environmental Research*, 193, 110520.

Stojić, A., Stanić, N., Vuković, G., Stanišić, S., Perišić, M., Šoštarić, A. and Lazić, L. (2019). Explainable extreme gradient boosting tree-based prediction of toluene,

ethylbenzene and xylene wet deposition. *Science of The Total Environment*, 653, 140-147.
Stojić, A., Stojić, S. S., Reljin, I., Čabarkapa, M., Šoštarić, A., Perišić, M. and Mijić, Z. (2016). Comprehensive analysis of PM_{10} in Belgrade urban area on the basis of long-term measurements. *Environmental Science and Pollution Research*, 23, 10722-10732.
Syuryavin, A. C., Park, S., Nirwono, M. M. and Lee, S. H. (2020). Indoor radon and thoron from building materials: Analysis of humidity, air exchange rate, and dose assessment. *Nuclear Engineering and Technology*, 52(10), 2370-2378.
Teixeira, E. C., Agudelo-Castañeda, D. M. and Mattiuzi, C. D. P. (2015). Contribution of polycyclic aromatic hydrocarbon (PAH) sources to the urban environment: a comparison of receptor models. *Science of the Total Environment*, 538, 212-219.
United States Environmental Protection Agency (US EPA) (1999). *Compendium of Methods for the Determination of Toxic Organic Compounds in Ambient Air. Determination of Polycyclic Aromatic Hydrocarbons (PAHs) in Ambient Air Using Gas Chromatography/Mass Spectrometry (GC/MS)*. Center for Environmental Research Information, Office of Research and Development U.S. Environmental Protection Agency, Cincinnati, USA. https://www.epa. gov/sites/production/files/ 2019-11/ documents/to-13arr.pdf. (Last accessed: February 4, 2021).
United States Environmental Protection Agency (US EPA) (2005). *Guidelines for carcinogen risk assessment*, EPA/630/P-03/001F, Washington, D.C., USA.
Varlet, V., Serot, T. and Prost, C. (2010). Smoke flavoring technology in seafood. *Handbook of seafood and seafood products analysis*, 233-254.
Vignet, C., Le Menach, K., Mazurais, D., Lucas, J., Perrichon, P., Le Bihanic, F., Devier, M. H., Lyphout, L., Frère, L., Bégout, M. L. and Zambonino-Infante, J. L. (2014). Chronic dietary exposure to pyrolytic and petrogenic mixtures of PAHs causes physiological disruption in zebrafish-part I: Survival and growth. *Environmental Science and Pollution Research*, 21, 13804-13817.
Vuong, Q. T., Thang, P. Q., Nguyen, T. N. T., Ohura, T. and Choi, S. D. (2020). Seasonal variation and gas/particle partitioning of atmospheric halogenated polycyclic aromatic hydrocarbons and the effects of meteorological conditions in Ulsan, South Korea. *Environmental Pollution*, 263, 114592.
Wang, J. and Ogawa, S. (2015). Effects of meteorological conditions on $PM_{2.5}$ concentrations in Nagasaki, Japan. *International Journal of Environmental Research and Public Health*, 12, 9089-9101.
Wick, A. F., Haus, N. W., Sukkariyah, B. F., Haering, K. C. and Daniels, W. L. (2011). *Remediation of PAH-contaminated soils and sediments: a literature review*. CSES Department, internal research document, 102.
Wilcke, W. (2000). Synopsis polycyclic aromatic hydrocarbons (PAHs) in soil—a review. *Journal of Plant Nutrition and Soil Science*, 163, 229-248.
Yang, X., Zhao, C., Guo, J. and Wang, Y. (2016). Intensification of aerosol pollution associated with its feedback with surface solar radiation and winds in Beijing. *Journal of Geophysical Research: Atmospheres*, 121, 4093-4099.

Yukhimets, A., Kuzu, S. L., Akyüz, E. and Saral, A. (2019). Investigation of geospatial distribution of PAH compounds in soil phase and determination of soil–air exchange direction in a megacity. *Environmental Geochemistry and Health*, 1-14.

Zelinkova, Z. and Wenzl, T. (2015). The occurrence of 16 EPA PAHs in food–a review. *Polycyclic Aromatic Compounds*, 35, 248-284.

Zhang, L., Morisaki, H., Wei, Y., Li, Z., Yang, L., Zhou, Q., Zhang, X., Xing, W., Hu, M., Shima, M. and Toriba, A. (2020). $PM_{2.5}$-bound polycyclic aromatic hydrocarbons and nitro-polycyclic aromatic hydrocarbons inside and outside a primary school classroom in Beijing: Concentration, composition, and inhalation cancer risk. *Science of the Total Environment*, 705, 135840.

Zhang, X. X., Cheng, S. P., Cheng-Jun, Z. H. U. and Shi-Lei, S. U. N. (2006). Microbial PAH-degradation in soil: degradation pathways and contributing factors. *Pedosphere*, 16, 555-565.

Zhao, L., Zhao, Y., Nan, H., Yang, F., Qiu, H., Xu, X. and Cao, X. (2020). Suppressed formation of polycyclic aromatic hydrocarbons (PAHs) during pyrolytic production of Fe-enriched composite biochar. *Journal of Hazardous Materials*, 382, 121033.

Zhao, Z., Zhang, L., Cai, Y. and Chen, Y. (2014). Distribution of polycyclic aromatic hydrocarbon (PAH) residues in several tissues of edible fishes from the largest freshwater lake in China, Poyang Lake, and associated human health risk assessment. *Ecotoxicology and Environmental Safety*, 104, 323-331.

Zhu, Y., Tao, S., Sun, J., Wang, X., Li, X., Tsang, D. C., Zhu, L., Shen, G., Huang, H., Cai, C. and Liu, W. (2019). Multimedia modeling of the PAH concentration and distribution in the Yangtze River Delta and human health risk assessment. *Science of the Total Environment*, 647, 962-972.

Živković, M. M., Jovašević-Stojanović, M., Cvetković, A., Lazović, I., Tasić, V., Stevanović, Ž. and Gržetić, I. A. (2015). PAHs levels in gas and particle-bound phase in schools at different locations in Serbia. *Chemical Industry and Chemical Engineering Quarterly/CICEQ*, 21, 159-167.

Chapter 2

Polycyclic Aromatic Hydrocarbon Contamination in Sharks and Batoids (Chondrichthyes: Elasmobranchii) and Ensuing Ecological Concerns

Natascha Wosnick[1], Mariana F. Martins[2], Gabriela A. V. Moog[3] and Rachel Ann Hauser-Davis[4],*

[1]Programa de Pós-Graduação em Zoologia,
Universidade Federal do Paraná (UFPR), PR, Brazil
[2]Programa de Pós-Graduação em Ciências Fisiológicas,
Universidade Federal do Rio Grande (FURG), Rio Grande, RS, Brazil
[3]Departamento de Ciências Biológicas,
Universidade Santa Úrsula (USU), Rio de Janeiro, RJ, Brazil
[4]Laboratório de Avaliação e Promoção da Saúde Ambiental,
Instituto Oswaldo Cruz (Fiocruz), Rio de Janeiro, RJ, Brazil

Abstract

Polycyclic Aromatic Hydrocarbons (PAHs) are ubiquitous organic compounds in the marine environment, originating mainly from anthropogenic sources. Due to their physico-chemical properties and ability to bioaccumulate in living organisms, PAHs are of great ecotoxicological concern, and their carcinogenic, and genotoxic properties make these compounds particularly harmful to exposed organisms. In this context, marine fauna assessments are paramount to estimate PAH environmental bioavailability, biochemical effects, as well

---

* Corresponding Author's Email: rachel.hauser.davis@gmail.com.

In: Polycyclic Aromatic Hydrocarbons
Editor: Warren L. Gregoire
ISBN: 978-1-68507-626-9
© 2022 Nova Science Publishers, Inc.

as potential ecological effects. Evaluations in marine organisms, such as bony fish, are readily available. Studies concerning sharks and batoids (elasmobranchs), however, are still scarce, which is interesting as this group is an important part of marine trophic network interactions and highly threatened by anthropogenic activities, including chemical contamination. In this context, this chapter will discuss PAH exposure and their associated effects in elasmobranchs. Furthermore, ensuing ecological concerns and public health implications due to contaminated elasmobranch meat consumption shall also be reflected upon.

Keywords: contaminants, ecotoxicology, elasmobranchs, marine, public health

1. Introduction

Polycyclic Aromatic Hydrocarbons (PAHs) comprise a class of organic pollutants comprising two or more aromatic rings. Compared to other organic compounds, PAHs are moderately lipophilic, with octanol water-partition coefficients (log KOW) ranging between 3.37 (naphthalene) and 6.75 (dibenzo[a,h]anthracene) (Latimer and Zeng, 2003). These compounds originate from both natural and anthropogenic processes, mainly from fossil fuel (gasoline, kerosene, diesel oil) and tar burning, as an industrial processing by-product (petrochemical effluents, refinery waste, industry vessels), forest fires, industrial waste incineration, cigarette smoke and tobacco, among others (Law & Biscaya, 1994). PAHs are termed petrogenic when originated from petroleum, i.e., naphthalene, phenanthrene and their alkylated derivatives and pyrogenic when originated from combustion processes (Neff, 1979). The latter are more resistant to biodegradation due to their strong interaction with consolidated particulate material (Bouloubassi and Salliot, 1993).

Due to their environmentally persistence, resistance to chemical and biological degradation and ability to bioaccumulate in living organisms from different trophic levels (Neves, 2002), PAHs comprise some of the most alarming organic contaminants (Honda and Suzuki, 2020). Unlike other anthropogenic contaminants banned in the past due to their persistence and ecotoxicological effects, PAHs are still chronically released into the environment, mostly due to fossil fuel combustion and petroleum production (Burgess et al., 2003). In this context, anthropogenic sources contribute considerably to the environmental input of these compounds, despite their biogenic natural production (Hylland, 2006).

Specific PAH effects in humans and other organisms are difficult to assess since these compounds are usually presented as mixtures (Baali and Yahyaoui, 2019). Despite this, genotoxic effects have been extensively associated to PAH exposure in a variety of animal models, leading to carcinogenicity and mutagenicity (Payne et al., 1989). The International Agency for Cancer Research, for example, has noted an increased risk of cancer following exposure to at least 60 different PAHs and materials containing these compounds (IARC, 2010) and the United States Environmental Protection Agency (EPA) lists several of these compounds as national priority pollutants, which must be frequently monitored in industrial effluents (EPA, 2008). Additionally, evaluating PAHs exposure and marine environment and biota effects is also of particular interest due to their association with oil spill events, since they constitute considerable portions of crude oil (Hayakawa et al., 2018).

2. Polycyclic Aromatic Hydrocarbons in the Marine Environment

Following atmospheric deposition and industrial and domestic effluents surface and underground runoff inputs, PAHs enter the marine environment and are distributed through the water column and sediments. When these compounds are introduced into the marine environment, they undergo partitioning, where low molecular weight compounds, with higher vapor pressures, tend to volatilize (Haritash and Kaushik, 2009), while less volatile and more soluble PAHs are absorbed onto the particulate material and deposited in the sediment (Cullen et al., 1994), where they can undergo chemical and biochemical changes. The persistence of these compounds in the aquatic environment also varies with their molecular mass, with lighter compounds containing fewer aromatic rings being degraded more easily (half-life in sediment of 9 and 43 days for naphthalene and anthracene, for example), while heavier PAH are more persistent (CETESB, 2021). Therefore, low molecular weight PAHs (LMW-PAHs) are more quickly absorbed by biota than adsorbed in sediments and are, thus, considered more toxic to marine organisms. At high latitudes, these contaminants tend to be immobilized due to low temperatures, although they can still be redistributed by physical transport processes, such as currents and waves (Kennicutt et al., 1995).

As PAHs are hydrophobic, their presence in marine ecosystems occurs due to their low solubility in water. Because of this, they quickly become associated with sediments, where they can persist until their degradation, resuspension, or bioaccumulation (Neves, 2002). On the other hand, several chemical or biochemical processes can transform these molecules into more reactive and, consequently, more toxic products. The incorporation of oxygen into the HPH, for example, leads to a polarizing effect, resulting in the production of reactive species, such as carbon dioxide. This is the reason why PAHs become more toxic to fish, for example, after exposure to ultraviolet radiation (Abdel-Shafy and Mansour, 2016). Volatile PAHs, although they remain in the environment for less time than higher molecular weight PAHs, are also potentially toxic.

PAHs released from the marine environment can be solubilized or adsorbed to suspended material and sediment, being consequently absorbed by marine organisms through water exposure, sediment ingestion or the dietary route. The main mechanisms involved in PAH absorption by aquatic biota are bioconcentration, bioaccumulation and biomagnification processes. Bioconcentration occurs through the absorption of PAHs through the skin or through gill ventilation, resulting in a higher concentration of these compounds in the body than in the surrounding water environment. Bioaccumulation comprises PAH absorption by organisms at higher rates than excretion, resulting in an increase in PAH concentrations at the organismic level. Finally, biomagnification processes comprise increases in contaminant concentrations along the food chain, as a function of trophic position (Mackay and Fraser, 2000).

In vertebrates such as fish, birds, and mammals, most absorbed PAHs are readily biotransformed and excreted (Varanasi et al., 1989), while in some invertebrates the metabolic capacity concerning these contaminants is lower (Walker, 2001). Additionally, PAHs can bioaccumulate in low trophic levels due to lower metabolization capabilities, while organisms belonging to higher trophic levels seem to readily biotransform these compounds (Eisler, 1987). Due to their lipophilic character, PAHs easily enter cells, allowing for differential tissue accumulation (Albers, 1995), and, in this regard, fatty tissues, such as mammary gland and adipose tissue, are considered significant storage deposits for these contaminants (Modica et al., 1983). Even so, PAHs can intervene in fertilization and reproduction (Mazurová et al., 2008), also causing immunological (Payne & Fancey, 1989) and endocrine (Mazurová et al., 2008) effects. At high concentrations, acute toxicity, both for fish and benthic invertebrates (Bellas et al., 2008; Mäenpää et al., 2009) may occur, and in fish, PAH can accumulate in eggs, being transferred to the larvae after hatching (Goksøyr et al., 1991). Despite inherent metabolic capacities, PAH concentrations in aquatic organisms can be influenced by other factors, such as water temperature, nutritional quality, stress, oxygen consumption and physiology (Lemke and Kennedy, 1997; Perugini et al., 2007).

PAH exposure in aquatic ecosystems is usually evaluated through the detection of these compounds in aquatic biota in tissues or biological fluids (Watson et al., 2004). Among the various organisms that have been proposed for use in biomonitoring in the marine environment, fish are noteworthy, as they play an essential role in the food chain and represent an important source of food for humans. However, the efficiency of assessing PAH concentrations in fish tissues has been questioned, considering that fish convert up to 99% of PAHs into metabolites within 24 hours of absorption, altering their pattern and concentrations in different tissues (Varanasi et al., 1989), due to a well-developed oxidase system, in which alkylated PAHs are rapidly metabolized, and most are secreted into bile and stored in the gallbladder before being excreted (Au et al., 1999). Because of this, PAH half-lives in fish are generally very short, for example from six to nine days for fluorene, phenanthrene, anthracene and fluoranthene (Meador et al., 1995). Therefore, an adequate alternative for these assessments is the analysis of PAH and their metabolites in bile, an interesting chronic and recent PAH contamination indicator, due to quick PAH metabolization (Lima, 2001).

3. Polycyclic Aromatic Hydrocarbon Contamination in Elasmobranchs

It is estimated that 36% of all Chondrichthyes (i.e., elasmobranchs and chimaeras) are threatened with extinction (IUCN, 2021), currently one of the most threatened taxonomic groups. Sharks (Elasmobranchii: Selachii) are cartilaginous fish with an evolutionary history dating back 400 million years, with species adapted to the most diverse environmental conditions. To date, 85 shark species are listed as Vulnerable (VU), 74 as Endangered (EN), and 55 as Critically Endangered (CR) (IUCN, 2021). Batoids (Elasmobranchii: Batoidea) consist of rays, skates, guitarfishes and sawfishes (Achliman et al., 2012) and represent over 50% of elasmobranch diversity. Currently, 101 batoid species are listed as VU, 64 as EN, and 63 as CR (IUCN, 2021), indicating that this group is more vulnerable to anthropogenic activities than previously thought. Population reduction by pollutants is a component for both global and regional risk assessments (i.e., IUCN - criteria A2e). Moreover, pollution has been recently identified as a priority research theme within Conservation Physiology (Cooke et al., 2021). Despite this, to date, scarce efforts have been made to elucidate the deleterious effects of contamination on elasmobranch survival and population recruitment. In fact, environmental pollution is only addressed in 4% of all published articles concerning sharks and batoids (Consales and Marsili, 2021), highlighting the urgent need to incorporate this subject to elasmobranch research.

PAHs elasmobranch exposure and effects have been barely assessed when compared to other contaminant classes, such as organochlorine pesticides and metals. Despite this, as regarding other contaminants, sharks and batoids accumulate considerable concentrations of chemical pollutants, especially in the liver, which is the main lipid storage organ in this group (Gelsleichter and Walker, 2010). Most chemical contaminant burdens are attributed to their high trophic position since most species act as meso- or apex predators. Furthermore, specific biological traits, such as slow growth, late maturation, and low fecundity, make this group particularly vulnerable to environmental stressors (García et al., 2008; Dulvy et al., 2014).

4. Polycyclic Aromatic Hydrocarbon Levels in Elasmobranchs Worldwide

Regarding PAH concentrations reported for elasmobranchs worldwide (Table 1), most studies report the total sum of PAHs ($\sum$PAHs) but different methodologies must be taken into account when comparing species and regions, since different parental compounds are considered when summing PAHs. Whereas most studies analyzing tissue concentrations report the 16 recommended priority PAHs, only some have determined methylated metabolites (Martins et al., 2020; Martins et al., 2021). Hepatic concentrations are the most reported for elasmobranchs, due to this organ's lipid-rich content combined to an important role in xenobiotic metabolism. Muscle is also commonly assessed in ecotoxicological studies, possibly due to its importance for human consumption. Bile is also a preferable matrix for assessing PAHs levels, through the assessment of fluorescent aromatic compounds (FACs). Gill and assessments are also available, but to a lesser extent (e.g., Al-Hassan et al., 2000). Finally, reports on ovary PAH concentration are restricted, even though they provide valuable inferences on reproductive and embryonic exposure scenarios (Martins et al., 2020). Although small-sized sharks are capable of assimilating PAHs, there is growing evidence that Carcharhiniformes and Lamniformes exhibit higher PAH concentrations. Moreover, most studies evaluating tissue burdens are concentrated in sharks belonging to the Carcharhiniformes order. Regarding batoids, guitarfishes (order Rhinopristiformes) seem to be the most studied group considering PAHs concentrations.

Table 1. Polycyclic aromatic hydrocarbon concentrations reported in different elasmobranch tissues worldwide. Concentrations are reported as µg g^{-1} wet weight in most studies

Species	Study site	Tissue	Concentration	Reference
Carcharhinus limbatus	Arabian Gulf	Gills	0.63 - 1.71	Al-Hassan et al. (2000)
		Liver	0.33 - 6.60	
		Muscle	0.15 - 5.26	
		Total	1.91 - 11.87	
Chylosyllim arabicum		Gills	1,00	
		Liver	5,78	
		Muscle	1,15	
		Total	7,93	
Chylosyllim arabicum		Gills	2,83	
		Liver	0,13	
		Muscle	n.d.	
		Total	2,97	
Carcharhinus sorrah		Gills	38,44	
		Liver	33,46	
		Muscle	1,02	
		Total	72,96	
Carcharhinus melanopterus		Gills	1,17	
		Liver	2,00	
		Muscle	1,66	
		Total	4,84	
Carcharhinus dussumieri		Gills	0,92	
		Liver	3,06	
		Muscle	0,71	
		Total	4,69	
Rhizoprionodon acutus		Gills	3.03 - 4.50	
		Liver	12.43 - 26.62	
		Muscle	2.68 - 34.84	
		Total	20.27 - 60.02	

Table 1. (Continued)

Species	Study site	Tissue	Concentration	Reference
Carcharodon carcharias	South Africa	Muscle	5.01 - 6.33*	Marsili et al. (2016)
Prioace glauca	South Africa		0.508 ± 0.027	Chukwumalume (2016)
Galeorhinus galeus			0.458 ± 0.013	
Isurus oxyrinchus			0.489 ± 0.016	
Mustelus mustelus			0.113 ± 0.04	
Mustelus schmitti	Argentina	Muscle	0.27 - 0.42	Oliva et al. (2017)
Carcarhinus leucas	Gulf of Mexico	Liver	1.56 ± 0.20	Cullen et al. (2019)
		Muscle	1.33 ± 0.16	
Carcarhinus limbatus		Liver	2.12 ± 0.11	
		Muscle	1.15 ± 0.08	
Sphyrna tiburo		Liver	2.22 ± 0.22	
		Muscle	1.18 ± 0.04	
Hexanchus griseus	Central Mediterranean Sea	Liver	<L.O.Q.	Salvo et al. (2019)
Carcarhinus leucas	USA inland	Liver	0.904 ± 0.11	Hernout et al. (2020)
Pseudobatos horkelii	Brazil	Blood	0.83 ± 0.54	Martins et al. (2020)
		Gills	1.56 ±1.89	
		Liver	1.45 ± 0.71	
		Muscle	2.134 ± 0.85	
		Ovaries	1.286 ± 0.47	

*Concentration reported as µg g^{-1} dry weight. LOQ – Limit of quantification. N.d.- not detected.

For sharks, the first paper published concerning exposure in wild specimens' dates from 2000, comprising PAH assessments in the gills, liver, and muscle tissue of six sharks from the Arabic Gulf, with $\sum$PAH concentrations ranging from 0.13 (Arabian carpet shark, *Chiloscyllium arabicum*) to 72.96 µg g^{-1} w.w. (Spot-tail shark, *Carcharhinus sorrah*), indicating levels reflecting the hazard of Arabian Gulf waters (Al-Hassan et al., 2000). This was also the first evidence of maternal PAH transfer in sharks (Blacktip shark, *Carcharhinus limbatus*), with similar concentrations in a pregnant female (2.07 µg g^{-1} ww) and her unborn neonates (2.17 µg g^{-1} w.w.). According to the authors, two- to four-ringed compounds (naphthalene, phenanthrene, fluoranthene, and pyrene) were predominant, possibly due to their higher solubility in water.

In a study performed in South Africa, PAHs levels were assessed in the Blue shark (*Prionace glauca*), Soupfin shark (*Galeorhinus galeus*), Shortfin mako shark (*Isurus oxyrinchus*), and Smooth-hound shark (*Mustelus*

mustelus), with total PAH concentrations of 0.508 ± 0.027 μg g^{-1} w.w., 0.458 ± 0.013 μg g^{-1} w.w., 0.489 ± 0.016. μg g^{-1} w.w., and 0.113 ± 0.04 μg g^{-1} w.w.), respectively (Chukwumalume, 2016). Moreover, a prevalence of naphthalene was detected, with strong evidence of petrogenic origin. Sharks from Argentina were also evaluated, with mean muscle PAH concentrations ranging from 0.27 to 0.42 μg g^{-1} w.w. in juvenile and adult narrow nose Smooth-hound sharks (*M. schmitti*) (Oliva et al., 2017). Again, low molecular weight PAHs (LMW-PAHs) represented most of the detected compounds, with naphthalene corresponding to the major portion of the total sum.

In Carcharhiniformes from Galveston Bay (Texas, Gulf of Mexico, USA), total hepatic PAH concentrations were significantly higher in Blacktip (2.1 μg g^{-1} w.w.) and Bonnethead (*Sphyrna tiburo*, 2.2 μg g^{-1} w.w.) sharks compared to Bull sharks (*C. leucas*, 1.6 μg g^{-1} w.w.), although no significant differences for muscle tissue were detected among species (Cullen et al., 2019).

Non-lethal methods for tissue sampling were employed when assessing PAHs in apex predators, as in the case of the Great white shark (*Carcharodon carcharias*), where skin and muscle biopsies of fifteen white sharks from the South African coast were obtained (Marsili et al., 2016). Males contained the highest PAH levels, followed by females. The presence of carcinogenic PAHs represented, respectively, $7.89 \pm 0.85\%$ in males and $7.45 \pm 1.65\%$ in females. Regarding non-carcinogenic PAHs, naphthalene was the most prevalent compound, followed by acenaphthene, phenanthrene, fluorene, pyrene, fluoranthene, benzo[g, h, i]perylene, and anthracene. As white sharks are one of the most important cosmopolitan and epipelagic top predators, categorized as Vulnerable according to the IUCN (Rigby et al., 2018), Marsili et al. (2016) report skin/superficial muscle biopsies as a non-lethal and less invasive method to evaluate environmental contaminant burdens in these animals. Furthermore, white sharks seem to be highly exposed to PAH contamination in Africa, a developing country with many agricultural products and industrial activities, especially with regard to oil and oil products traffic, as almost 28% of oil exported in the Middle East passes near Cape of Good Hope (Moldan et al., 1987). The authors postulate the latter as being the probable cause for the high PAH values detected in the investigated white shark specimens.

Moreover, with the increased marine pollution noted at the global level, deep-sea species are also increasingly affected by pollutants, including PAHs. In a study conducted in the Mediterranean Sea, for example, hepatic PAH accumulation in the Bluntnose sixgill shark (*Hexanchus griseus*) was assessed, with a total of 13 PAHs (acenaphthylene, fluorene, phenanthrene, anthracene, pyrene, benzo[a]pyrene, chrysene, benzo[b]fluoranthene, benzo[k] fluoranthene, benzo[a]anthracene, indene[1,2,3-cd]pyrene, dibenzo[a,h] anthracene and benzo[g, h, i]perylene) identified, albeit below the limit of quantification (LOQ) of the applied analytical method (Salvo et al., 2019). Integrated analyses have also been carried out, aiming to analyze potential correlations between hepatic PAH body burdens and fatty acids in Bull sharks (Hernout et al. 2020). Results indicate that sharks exhibited the highest levels of HMW-PAHs ($\geq$ four aromatic rings) although no clear correlation with any particular type of fatty acid was noted. Furthermore, PAHs levels detected in at Sabine Lake (Texas, Gulf of Mexico, USA) were similar to those detected in *C. leucas* individuals sampled in Galveston Bay (110.39 km away), pointing to chronic exposure in both locations and Bull shark inability to eliminate PAHs at faster rates than their absorption (Cullen et al. 2019).

Martins et al. (2020) evaluated the tissue distribution of 18 PAHs in the Brazilian guitarfish (*Pseudobatos horkelii*) in the Southwest Atlantic, reporting mean concentrations ranging from 0.83 to 2.13 $\mu g \, g^{-1}$ w.w. in blood and muscle samples. The authors observed a clear distribution of PAHs, with higher h liver and muscle concentrations, followed by gills and ovaries. Contaminant partitioning in gonads was suggested as indicative of maternal transfer and further assessed in this species. A transfer rate of 13% of total sum of PAHs was calculated and physico-chemical properties (log KOW) of each PAH were associated with individual maternal transfer rates, resulting in an elevated transfer rate for lower log KOW compounds. A molecular weight influence in crossing biological membranes was suggested as the major factor for the high transference of these compounds. Finally, a trend was noted where larger females presented lower PAH concentrations and higher transfer rates, suggesting maternal PAH offloading as a potential depuration mechanism in guitarfishes.

PAH exposure determination through bile fluorescence was carried out for the Chilean catshark *Schroederichthys chilensis* as an effective and low coast method when compared to other chromatographic approaches (Fuentes-Rios et al., 2005). The authors reported higher fluorescence peaks in organisms collected in impacted areas, suggesting PAH exposure. Moreover, 1-OH-pyrene and naphthalene metabolites were pointed out as fluorescence-inducing metabolites. Finally, whereas 1-OH-pyrene occurrence was associated to more impacted sites, naphthalene metabolites were ubiquitous in the environment. Furthermore, biliary PAHs were also linked to xenobiotic metabolism induction.

5. Biochemical and Physiological Polycyclic Aromatic Hydrocarbon Effects in Elasmobranchs

Along with other biomarkers, cytochrome P450 (CYP1A) is inducible by PAHs, being traditionally employed in vertebrates concerning PAH biomonitoring efforts (Goldstone et al., 2006) and comprising the most assessed biomarker among elasmobranch studies. This enzyme was first detected for the Smooth dogfish (*Mustelus canis*), through immunoblotting performed on hepatic microsomes. Its inducibility was much higher in a dogfish captured within a polychlorinated biphenyl (PCB)-contaminated site, indicating its potential as a biomarker in sharks (Hahn et al. 1998).

CYP1A induction, measured as Ethoxyresorufin-O-deethylase (EROD) activity, has been observed in little skates (*Raja erinacea*), Atlantic stingrays (*Hypanus sabinus*) and dogfish sharks (*Squalus acanthias*) exposed to 3-Methylcolanthrene. Despite higher enzymatic activity, CYP14 transcription levels were not influenced, suggesting another cytochrome enzyme role in the xenobiotic metabolism of these species. Furthermore, benzo[a]pyrene hydroxylase was also induced, whereas glutathione-S-transferase (GST) activity was not affected (James and Bend, 1980).

In another study, performed with the Chilean catshark (*Schroederichthys chilensis*) off the Chilean coast (South Pacific Ocean), hepatic EROD activity associated with Fluorescent Aromatic Compounds (FACs) in shark bile were evaluated as potential biomarkers to access PAH exposure. Sharks sampled from highly impacted areas display the highest bile concentrations of 1-OH pyrene and benzo(a)pyrene-type metabolites (Fuentes-Rios et al. 2005), probably as reflecting local boat traffic and combustion-based industries, validating the use of both markers for sharks.

In 2010, a survey performed before and after the Deepwater Horizon oil spill off the coast of Alabama (USA) encompassing 14 shark species reported that CYP1A activity was not detected in any of sharks before the spill. However, about 3.4% of post-oil sampled sharks exhibited detectable EROD activity levels (Walker, 2011). Among the studied species, the Blacknose shark (*Carcharhinus acronotus*) was proven the best EROD induction indicator following oil exposure. Except for one Tiger shark (*Galeocerdo cuvier*) (16.75 pmol min^{-1} mg^{-1} protein), the observed EROD activities were similar to those observed by Fuentes-Rios et al. (2005), indicating that shark species may respond similarly to PAH exposure regarding EROD activity (Walker 2011). Still regarding the Deep Horizon Oil Spill, 11 shark species were assessed regarding the xenobiotic metabolism of CYP1A (EROD) and GST enzyme activities between reference and impacted areas. Comparisons between sites were possible for *C. niaukang* and *S. cf mitsukurii*, indicating the induction of both enzymes. Coastal species presented higher mean EROD (*Carcharhinus falciformis*, 1.78 pmol min^{-1} mg^{-1} protein) and GST (*Carcharhinus plumbeus*, 100.2 nmol/min/mg protein) activities, whereas deep dwelling species presented the lowest enzymatic activities for both EROD (*C. cf. granulosus*, 0.01 pmol min^{-1} mg^{-1}) or GST (*S. cf mitsukurii*, 29.7 nmol min^{-1} mg^{-1} protein) (Leary, 2015).

Biochemical assessments (i.e., EROD and GST activities) were also carried out in the sharks from Galveston Bay (Cullen et al., 2019), and only weak correlations with no possible explanations were noted. Dibenzo[a,h]anthracene and indeno[1,2,3, c-d]pyrene, however, displayed a positive relationship to EROD activity. Toxic equivalents (TEQ) were also calculated and observed to be higher in the species exhibiting the highest PAH burdens (*C. limbatus* and *S. tiburo*). These features did not correlate to EROD levels, possibly due to low PAH bioavailability in the liver (i.e., compounds could be deposited in lipid-rich compartments and not available for the enzymatic metabolism). The authors suggest potential physiological effects and accumulation of PAH burdens compared to established TEQ thresholds for other taxa, although they state that the likelihood of similar effects in sharks requires further studies and the inclusion of toxic endpoints. Furthermore, lipid peroxidation has also been observed in Carcharhiniformes sharks presenting relatively high hepatic PAH levels (Cullen et al., 2019). Lipid damage is highly associated with oxidative stress, which, in turn, might be triggered by a variety of factors, including EROD induction (Whyte et al., 2000; Van der Oost et al., 2003). EROD induction might also indirectly influence the formation of DNA adducts by transforming parental PAHs into metabolites, which may be more toxic (Varanasi et al., 1989; Billiard et al., 2008). Although most of these effects are constrained to low hierarchical levels, long term organismal and population-level effects derived from mild alterations should not be neglected.

For batoids, Whalen (2017) also assessed phase I and II biotransformation enzymes, observing that Atlantic stingrays (*Hypanus sabinuns*) displayed different EROD and GST activities, as well as lipid peroxidation (LPO) levels as a function of sampling site, with increased biomarkers in more impacted areas. The author also employed an Integrated Biomarker Response analysis (IBR), which was higher for the most impacted site, suggesting decreased ecosystems health. Such an approach indicates the possibility of using elasmobranchs in multi-biomarker analyses beyond individual or specific aims.

Interesting and unusual approaches have been evaluated using the Atlantic Stingray (*Hypanus sabinuns*) in North America during the past years, where an integrated biomarker response approach (IBR) was applied to enzymatic and cell damage biomarkers linked to biliary PAH metabolites using this species to assess ecosystem health in Florida, USA (Whalen, 2017).

Systemic disruptions related to crude oil exposure were also studied in this same species (*Hypanus sabinuns*). Although not directly associated to PAH exposure, PAHs constitute a significant proportion of crude oil. Thus, olfactory responses were assessed through testing the electrophysiological activity to specific amino acids known to stimulate this species (Cave and Kajiura, 2018). Authors attribute the delayed response in exposed individuals to either physical or physiological contaminant effects in the olfactory organ, by diffusing molecules or increasing mucus production and, therefore, delaying molecule binding to cell receptors. Later, in 2020, acute crude oil exposure was linked to impaired olfactory and electrosensory functions in this species. Electrosensory function was also disrupted through a decreased electric field perception compromising the foraging ability of this species (Cave and Kajiura, 2020).

Some speculative associations have been proposed based on individuals captured in areas where environmental levels of PAHs are known *a priori*. However, correlations must be carefully addressed, since contaminants have not been determined in these organisms and the study site was not directly impacted by crude oil, as in other studies carried out in oil spills scenarios, (Walker, 2011; Leary, 2015; Whalen, 2017). Karsten and Rice (2004), for example, proposed PAH exposure as a potential factor influencing elevated c-reactive protein levels in sharpnose sharks (*Rhizoprionodon terranovae*). Despite this, other stressors have also been pointed out as possible causes for the observed proinflammatory responses.

6. Potential Ecological Outcomes of Polycyclic Aromatic Hydrocarbon Contamination in Elasmobranchs

Biochemical responses to chemical stressors such as contaminants are key in identifying future irreversible effects of environmental pollution, and biomarkers are highly applied for this purpose in wild populations (Van der

Oost et al., 2003). Elasmobranchs display enzymatic responses associated with PAH exposure, assessed through organism PAH burden or occurrence in contaminated areas, especially regarding phase I detoxification enzymatic activity (represented by EROD activity). In most studies, increased EROD activity was associated with impacted areas (Walker, 2011; Whalen, 2017) or exposed organisms (Fuentes-Rios et al., 2005), indicating the activation of the xenobiotic biotransformation mechanism as a contaminant exposure factor in these animals. This, in turn, leads to involuntary formation of reactive oxygen species (ROS) which, consequently, result in detrimental cellular oxidative imbalance and further cell constituent damage (Luschack, 2015). Outcomes, are however, still unknown regarding long-term survival and negative n systemic health and individual resilience and fitness effects.

Biomarkers of effect *per se*, such as lipid peroxidation, have less observed in elasmobranchs, but are probably more due to the lack of studies using this approach rather than a lack of observations in this context. However, enzymatic induction triggered by the xenobiotic metabolism demands energy, which might lead to energy imbalances and sublethal effects at the organismic level, affecting growth and reproduction, for example (Moiseenko, 2010). Behavioral alterations, led by sensory cell impairment or crude oil exposure are also potential outcomes with direct ecological impacts, as the impairment of sensory systems has been previously observed in stingrays acutely exposed to crude oil containing a set of contaminants, including PAHs, from the Deep Horizon Oil Spill (Cave and Kajiura, 2018; 2020). In addition, enzymatic inhibition and other biochemical contaminant effects, such as cell damage or death, leading to a decreased number of olfactory receptors are also considered potential variables. However, it is important to note that, despite the relevant contributions of these experimental designs, interpretations must be carefully linked to PAHs only, since crude oil is comprised of mixture of compounds. All of the aforementioned alterations have a direct impact on foraging and survival, potentially influencing several species at the population level in impacted areas.

Evidence of embryonic exposure to PAHs in elasmobranchs is concerning, considering that early life stages are more vulnerable to environmental contamination (Jezierska et al., 2009), especially due to lower biotransformation capacities, compromising xenobiotic detoxification abilities (Russel et al., 1999). Some studies have reported this phenomenon in sharks and batoids (Al-Hassan et al., 2000; Martins et al., 2021), with considerable concentrations observed in offspring. Previous studies have associated PAH exposure with embryonic abnormalities in fishes (Incardona et al., 2004) and survival failure in birds (Brunström et al., 1990). Given the conservative physiology of vertebrates, one might expect that such consequences also take place in elasmobranchs exposed to similar concentrations or for longer periods of time.

7. Public Health Implications of Polycyclic Aromatic Hydrocarbon Contamination in Elasmobranchs

Problems arising from the toxic effects of PAHs on aquatic ecosystems are not restricted to ecological imbalances but can also affect human health in the long term, if the possibility of the occurrence of toxic pollutant persistence and bioaccumulation is considered along the food chain (Houk, 1992). Beyond their ecological role in ecosystems, elasmobranchs have become a supplementary resource in fisheries, mostly due to overexploitation. Despite the well-known Asian role concerning shark fin imports and consumption, the major elasmobranch meat consumers (hereafter referred as "shark" meat) are located in South America (mostly Brazil and Uruguay) and Europe (represented by Italy) (Dent and Clark, 2015). Particularly in developing countries, shark meat is an important protein source (Bornatowski et al., 2018). The consumption of shark-derived products such as meat, oil and vertebrae, among others can pose risks to human health, where certain populations, such as pregnant women and young children, are especially vulnerable to exposure to these contaminants (Patandin et al., 1999; Bruce-Vanderpuije et al., 2019). In Brazil, for instance, shark meat is purchased and offered to students in public schools in Rio de Janeiro (SEEDUC 2019), highlighting the need for adequate food quality and safety monitoring regarding contaminant levels in edible shark tissues.

PAHs have been determined in a variety of food items and dietary exposure is an important contaminant route for these compounds in humans (Phillips, 1999). Effects associated with PAH exposure mostly comprise toxicity and cancer (Mastrangelo et al., 1996), but also include reproductive impairment and neurotoxicity (see Ramesh et al., 2011 for a review), among others. PAH risk characterizations are mostly assessed through the carcinogenicity and mutagenicity induction potential of these compounds. Only some countries have established their own legislation for PAH concentrations in food items. The European Commission, for example, sets the limits of 2.0 and 12.0 $\mu g\ kg^{-1}$ lipid weight for benzo[a]pyrene and the sum of benzo[a]pyrene, benzo[a]anthracene, benzo[a]fluoranthene and chrysene, respectively, concerning smoked fish meat (Zelinkova and Wenzl, 2015). Despite this, this regulation concerns smoked or heated seafood, and most studies evaluating PAHs in sharks use fresh muscle samples. Additionally, benzo[a]pyrene was recommended as a marker of PAH occurrence and carcinogenicity in foodstuffs by the Scientific Committee on Food (SCF) (EFSA, 2008). This compound has been determined in several shark species muscle samples, ranging from 3.4 to 36.7% in Bull sharks (Cullen et al., 2019) and Great white sharks (Marsili et al., 2016), respectively, indicating public health concerns.

Some studies have attempted to assess the health risks of consuming elasmobranchs meat but failed to detect any equivalents above established safety limits. Oliva et al. (2017), for example, assessed the health risk of consuming Narrow nose smooth-hound meat in the Argentinean population, through the assessment of toxic equivalents of benzo[a]pyrene (TEQ BaP) and daily dietary intakes (DDI). The authors detected a TEQ of 0.57 ng g^{-1}, below the suggested safety limit (USEPA, 2000). Additionally, Sun et al. (2019) calculated the carcinogenic risk (CRL) of PAHs in the Spiny skate (*Raja porosa*) (1.02 x 10^{-6}), and reported that, although a certain risk was observed, it was still less than recommended maximum permissible levels.

Conclusion

Although PAHs are rapidly metabolized by fishes, high burdens of these compounds, at the $\mu g\ g^{-1}$ level, have been detected in elasmobranchs. Hepatic and muscle tissues have been the most investigated so far in this group, so the toxicokinetics and tissue distribution of these compounds are still poorly understood. As PAHs have the potential to affect elasmobranch health and fitness, future studies should consider multi-tissue analyses, focusing on other important organs, such as the kidneys and gonads.

EROD enzymatic activity associated with PAH exposure is the most reported biomarker among elasmobranch species. Most studies report an induction in this enzyme in either impacted areas or exposed organisms. However, results should be carefully interpreted due to the complexity of contaminant mixtures present in the environment, which have led to combined effects on the xenobiotic metabolism. Considering the conservative physiology of vertebrates, future studies should focus on investigating other biomarkers already validated for other taxonomic groups.

Biomarkers of effect, such as lipid peroxidation, have been poorly assessed in studies evaluating PAHs effects on elasmobranchs, highlighting the urgent need to apply such approaches in ecotoxicological studies to predict cellular damage induced by environmental contamination. In addition, other damage biomarkers already employed to access the effects of other pollutants in elasmobranchs should be investigated, focusing on the impacts of PAHs on homeostatic balance and other systemic health metrics.

Despite the prevalence of reports concerning low hierarchical level effects (mostly at the biochemical level), systemic level effects have been reported for elasmobranchs, indicating that PAH effects may affect individual foraging and predation activities, and, therefore, survival. Thus, future studies should focus on a multidisciplinary approach, aiming to elucidate both ecological and behavioral PAH exposure outcomes through telemetry (field studies), and controlled experiments/ethograms (captive studies), among others.

The relatively high concentrations reported in edible tissues are critical considering that elasmobranchs are routinely consumed worldwide, especially in developing countries. Moreover, considering that most studied species are of commercial value, assessing PAH burdens and estimating food safety and monitoring metrics is also paramount regarding public health concerns.

References

Abdel-Shafy, H. I. & Mansour, M. S. M. (2016). "A review on polycyclic aromatic hydrocarbons: Source, environmental impact, effect on human health and remediation". *Egyptian Journal of Petroleum*, 25 (1), 107–123. https://doi.org//10.1016/j.ejpe.2015.03.011.

Al-Hassan, J. M., Afzal, M., Rao, C. V. N. & Fayad, S. (2000). "Petroleum hydrocarbon pollution in sharks in the Arabian Gulf". *Bulletin of Environmental Contamination and Toxicology*, 65, 391–398. https://doi.org/ 10.1007/s001280000140.

Albers, P. H. (1995). "Petroleum and individual polycyclic aromatic hydrocarbons". In: *Handbook of Ecotoxicology*, edited by D.J. Hoffman, B. A Rattner, G. A. Burton, J. Cairns. 330–355. Boca Raton: Lewis Publishers. https://doi.org/10.1201/9781420032505.ch14.

Au, D. W. T., Wu, R. S. S., Zhou, B. S. & Lam, P. K. S. (1999). "Relationship between ultrastructural changes and EROD activities in liver of fish exposed to benzo [a] pyrene". *Environmental Pollution*, 104 (2), 235–247.

Baali, A. & Yahyaoui, A. (2019). "Polycyclic Aromatic Hydrocarbons (PAHs) and their influence to some aquatic species". In: *Biochemical Toxicology - Heavy Metals and Nanomaterials*, edited by Ince, M., Ince, O.K., Ondrasek, G. London: IntechOpen. https://doi.org/10.5772/intechopen.86213.

Bellas, J., Fernández, N., Lorenzo, I. & Beiras, R. (2008). "Integrative assessment of coastal pollution in a Ría coastal system (Galicia, NW Spain): correspondence between sediment chemistry and toxicity". *Chemosphere*, *72*(5), 826–835. https://doi.org/10.1016/ j.chemosphere.2008.02.039.

Billiard, S. M., Meyer, J. N., Wassenberg, D. M., Hodson, P. V. & Di Giulio, R. T. (2008). "Nonadditive effects of PAHs on Early Vertebrate Development: Mechanisms and Implications for Risk Assessment". *Toxicological Sciences*, *105*, 5–23. https://doi.org/10.1093/toxsci/kfm303.

Bornatowski, H., Braga, R. R. & Barreto, R. P. (2018). Elasmobranchs consumption in Brazil: impacts and consequences. In: *Advances in Marine Vertebrate Research in Latin America* edited by Rossi-Santos, M. R., Finkl, C. W. New York: Springer International Publishing. https://doi.org/10.1007/ 978-3-319-56985-7_10.

Bouloubassi, I. & Saliot, A. (1993). "Investigation of anthropogenic and natural organic inputs in estuarine sediments using hydrocarbon markers (NAH, LAB,PAH) ". *Oceanologica Acta*, *16*, 145–161.

Bruce-Vanderpuije, P., Megson, D., Reiner, E. J., Bradley, L., Adu-Kumi, S. & Gardella Jr, J. A. (2019). "The state of POPs in Ghana-A review on persistent organic pollutants: Environmental and human exposure". *Environmental Pollution*, *245*, 331–342.

Brunström, B., Broman, D. & Näf, C. (1990). "Embryotoxicity of polycyclic aromatic hydrocarbons (PAHs) in three domestic avian species, and of PAHs and coplanar polychlorinated biphenyls (PCBs) in the common eider". *Environmental Pollution*, *67*, 133–143. https://doi.org/10.1016/0269-7491 (90)90078-Q.

Burgess, R. M., Ahrens, M. J. & Hickey, C. W. (2003). Geochemistry of PAHs in Aquatic Environments: Source, Persistence and Distribution. In: *PAHs: An Ecotoxicological Perspective*, edited by Douben, P.E.T. New York: Wiley. https://doi.org/10.1002/ 0470867132.ch3.

Cave, E. J. & Kajiura, S. M. (2018). "Effect of Deepwater Horizon crude oil water accommodated fraction on olfactory function in the Atlantic Stingray, *Hypanus sabinus*". *Scientific Reports*, *8*, 15786. https://doi.org/10.1038/ s41598-018-34140-0.

Cave, E. J. & Kajiura, S. M. (2020). "Electrosensory impairment in the Atlantic Stingray, *Hypanus sabinus*, after crude oil exposure". *Zoology*, *143*, 125844. https://doi.org/ 10.1016/j.zool.2020.125844.

CETESB. (2021). Available at: http://www.cetesb.sp.gov.br.

Chukwumalume, R. M. C. (2016). *Polycyclic aromatic hydrocarbons (PAHs) and organochlorine pesticide residues in selected marine fish species along the coast of South Africa*. PhD dissertation, Stellenbosch University. Available at http://scholar.sun.ac.za/handle/ 10019.1/98429.

Consales, G. & Marsili, L. (2021). "Assessment of the conservation status of Chondrichthyans: underestimation of the pollution threat". *The European Zoological Journal*, *88* (1), 165–180. https://doi.org/10.1080/24750263. 2020.1858981.

Cooke, S. J., et al. (2021). "One hundred research questions in conservation physiology for generating actionable evidence to inform conservation policy and practice". *Conservation Physiology*, *9* (1), coab009. https://doi.org/ 10.1093/conphys/coab009.

Cullen, W. R., Li, X. F. & Reimer, K. J. (1994). "Degradation of phenanthrene and pyrene by microorganisms isolated from marine sediment and seawater". *Science of the Total Environment*, *156* (1), 27–37.

Cullen, J. A., Marshall, C. D. & Hala, D. (2019). "Integration of multi-tissue PAH and PCB burdens with biomarker activity in three coastal shark species from the northwestern Gulf of Mexico". *Science of the Total Environment*, *650* (1), 1158–1172. https://doi.org/10.1016/j.scitotenv.2018.09.128.

Dent, F. & Clarke, S. (2015). "State of the global market for shark products". *FAO Fisheries and Aquaculture technical paper no. 590*. FAO, Rome

Dulvy, N. K., Fowler, S. L., Musick, J. A., Cavanagh, R. D., Kyne, P. M., Harrison, L. R., Carlson, J. K., Davidson, L. N. K., Fordham, S. V., et al. (2014). "Extinction risk and conservation of the world's sharks and rays". *eLife*, *3*, e00590. https://doi.org/10.7554/eLife.00590.001.

EFSA. (2008). "Polycyclic Aromatic Hydrocarbons in Food 1 Scientific Opinion of the Panel on Contaminants in the Food Chain Adopted on 9 June 2008". *European Food Safety Authority Journal*, *724*, 1–114.

Eisler, R. (1987). *"Polycyclic aromatic hydrocarbon hazards to fish, wildlife, and invertebrates: a synoptic review* (No. 11)". Fish and Wildlife Service, US Department of the Interior.

EPA. (2008). *"Polycyclic Aromatic Hydrocarbons (PAHs)"*. Available at: https://archive.epa.gov/epawaste/hazard/wastemin/web/pdf/pahs.pdf.

Fuentes-Rios, D., Orrego, R., Rudolph, A., Mendoza, G., Gavilán, J. F. & Barra, R. (2005). "EROD activity and biliary fluorescence in *Schroederichthys chilensis* (Guichenot 1848): Biomarkers of PAH exposure in coastal environments of the South Pacific Ocean". *Chemosphere*, *61*, 192–199.

García, V. B., Lucifora, L. O. & Myers, R. A. (2008). "The importance of habitat and life history to extinction risk in sharks, skates, rays and chimaeras". *Proceedings of the Royal Society B*, *275*, 83–89. https://doi.org/10.1098/rspb.2007.1295.

Gelsleichter, J. & Walker, C. J. (2010). "Pollutant exposure and effects in sharks and their relatives". In: *Sharks and Their Relatives II: Biodiversity, Adaptive Physiology, and Conservation*, edited by Carrier, J.C., Musick, J.A., Heithaus, M.R. Boca Raton: CRC Press.

Goksøyr, A., Husøy, A. M., Larsen, H. E., Klungsøyr, J., Wilhelmsen, S., Maage, A., Brevik, E. M., Andersson, T., Celander, M., Pesonen, M. & Förlin, L. (1991). "Environmental contaminants and biochemical responses in flatfish from the Hvaler Archipelago in Norway". Archives of *Environmental Contamination and Toxicology*, *21*(4), 486–496. https://doi.org/10.1007/ BF01183869.

Goldstone, H. M. & Stegeman, J. J. (2006). "A revised evolutionary history of the CYP1A subfamily: gene duplication, gene conversion, and positive selection". *Journal of Molecular Evolution*, *62* (6), 708–717. https://doi.org/ 10.1007/s00239-005-0134-z.

Hahn, M. E., Woodin, B. R., Stegeman, J. J. & Tillitt, D. E. (1998). "Aryl hydrocarbon receptor function in early vertebrates: Inducibility of cytochrome P450 1A in agnathan and elasmobranch fish". *Comparative Biochemistry and Physiology Part C: Pharmacology, Toxicology and Endocrinology*, *120* (1), 67–75. https://doi.org/10.1016/s0742-8413(98) 00007-3.

Haritash, A. K. & Kaushik., C. P. (2009). "Biodegradation aspects of Polycyclic Aromatic Hydrocarbons (PAHs): A review". *Journal of Hazardous Materials*, *169* (1–3), 1–15. https://doi.org/10.1016/j.jhazmat.2009.03.137.

Hayakawa, K. (2018). Oil spills and Polycyclic Aromatic Hydrocarbons. In: *Polycyclic Aromatic Hydrocarbons*, edited by Hayakawa K. Singapore: Springer, https://doi.org/10.1007/978-981-10-6775-4_16.

Hernout, B., Leleux, J., Lynch, J., Ramaswamy, K., Faulkner, P., Matich, P. & Hala, D. (2020). "The integration of fatty acid biomarkers of trophic ecology with pollutant body-burdens of PAHs and PCBs in four species of fish from Sabine Lake, Texas". *Environmental Advances*, *1*, 100001. https://doi.org/ 10.1016/j.envadv.2020.100001.

Honda, M. & Suzuki, N. (2020). "Toxicities of Polycyclic Aromatic Hydrocarbons for Aquatic Animals". *International Journal of Environmental Research and Public Health*, *17* (4), 1363. https://doi.org/10.3390/ijerph1704 1363.

Houk, V. S. (1992). "The genotoxicity of industrial wastes and effluents: a review". *Mutation Research/Reviews in Genetic Toxicology*, *277* (2), 91–138.

Hylland, K. (2006). "Polycyclic aromatic hydrocarbon (PAH) ecotoxicology in marine ecosystems". *Journal of Toxicology and Environmental Health Part A.*, *69*, 109e123. https://doi.org/ 10.1080/15287390500259327.

IARC Working Group on the Evaluation of Carcinogenic Risks to Humans. (2010). Some non-heterocyclic polycyclic aromatic hydrocarbons and some related exposures. *IARC IARC monographs on the evaluation of carcinogenic risks to humans.*, *92*, 1–853.

Incardona, J. P., Collier, T. K. & Scholz, N. L. (2004). "Defects in cardiac function precede morphological abnormalities in fish embryos exposed to polycyclic aromatic hydrocarbons". *Toxicology and Applied Pharmacology.*, *196*, 191–205. https://doi.org/10.1016/j.taap.2003.11.026.

IUCN. (2021). *The IUCN red list of threatened species*. Avaiable at: www.iucnredlist.org.

James, M. O. & Bend, J. R. (1980). "Polycyclic aromatic hydrocarbon induction of cytochrome p-450-dependent mixed-function oxidases in marine fish". *Toxicology and Applied Pharmacology*, *54*, 117–133. https://doi.org/ 10.1016/0041-008X(80) 90012-5.

Jezierska, B., Lugowsa, K. & Witeska, M. (2009). "The effects of heavy metals on embryonic development of fish (a review)". *Fish Physiology and Biochemistry*, *35*, 625–640. https://doi.org/10.1007/s10695-008-9284-4.

Karsten, A. H. & Rice, C. D. (2004). "c-Reactive protein levels as a biomarker of inflammation and stress in the Atlantic sharpnose shark (*Rhizoprionodon terraenovae*) from three southeastern USA estuaries". *Marine Environmental Research*, *58*, 747–751. https://doi.org/10.1016/j.marenvres.2004.03.089.

Kennicutt, II, M. C., McDonald, S. J., Sericano, J. L., Boothe, P., Oliver, J., Safe, S., Presley, B. L., Liu, H., Wolfe, D., Wade, T. L., Crockett, A. & Bockus, D. (1995). "Human contamination of the marine environment-Arthur Harbor and McMurdo Sound, Antarctica". *Environmental Science & Technology*, *29* (5), 1279–1287. https://doi.org/10.1021/es00005a600.

Latimer, J. S. & Zenger, J. (2003). The Sources, Transport, and Fate of PAHs in the Marine Environment. In: *PAHs: An Ecotoxicological Perspective*. Douben, P.E.T. New York: Wiley. https://doi.org/10.1002/0470867132.ch2.

Law, R. J. & Biscaya, J. L. (1994). "Polycyclic Aromatic Hydrocarbons (PAH) - Problems and progress in sampling analysis and interpretation". *Marine Pollution Bulletin*, 29, 235–241. https://doi.org/10.1016/0025-326X(94)90415-4.

Leary, A. E. (2015). *"Effects of the Deepwater Horizon Oil Spill on Deep Sea Fishes"*. Graduate Theses and dissertations. University of North Florida (UNF). Available at https://digitalcommons.unf.edu/cgi/ viewcontent.cgi?article=1596&context=etd.

Lemke, M. A. & Kennedy, C. J. (1997). "The uptake, distribution and metabolism of benzo[a]pyrene in coho salmon (*Oncorhynchus kisutch*) during the parr-smolt transformation." *Environmental Toxicology and Chemistry*, 16(7), 1384–1388. https://doi.org/10.1002/etc.5620160708.

Lima, E. F. (2001). *Acumulação de hidrocarbonetos policíclicos aromáticos e metais traço em invertebrados marinhos e avaliação do uso de biomarcadores celulares e bioquímicos no biomonitoramento* [Accumulation of polycyclic aromatic hydrocarbons and trace metals in marine invertebrates and evaluation of the use of cellular and biochemical biomarkers in biomonitoring]. PhD thesis, Pontifícia Universidade Católica do Rio de Janeiro (PUC-Rio).

Luschack, V. I. (2015). "Environmentally induced oxidative stress in aquatic animals". *Aquatic Toxicology*, 101, 13–30. https://doi.org/10.1016/j.aquatox.2010.10.006.

Mäenpää, K., Leppänen, M. T. & Kukkonen, J. V. K. (2009). "Sublethal toxicity and biotransformation of pyrene in *Lumbriculus variegatus* (Oligochaeta)". *Science of the total environment*, 407(8), 2666–2672. https://doi.org/ 10.1016/j.scitotenv.2009.01.019.

Mackay, D. & Fraser, A. (2000). "Bioaccumulation of persistent organic chemicals: mechanisms and models." *Environmental pollution*, 110(3), 375–391. https://doi.org/10.1016/S0269-7491(00)00162-7.

Marsili, L., Coppola, D., Giannetti, M., et al. (2016). "Skin biopsies as a sensitive non-lethal technique for the ecotoxicological studies of Great White Shark (*Carcharodon carcharias*) sampled in South Africa. " *Journal of Expert Opinion on Environmental Biology*, 5, 1 https://doi.org/10.4172/2325-9655. 1000126.

Martins, M. F., Costa, P. G. & Bianchini, A. (2020). "Contaminant screening and tissue distribution in the critically endangered Brazilian guitarfish *Pseudobatos horkelii*". *Environmental Pollution*, 265, 114923. https://doi. org/10.1016/j.envpol.2020.114923.

Martins, M. F., Costa, P. G. & Bianchini, A. (2021). "Maternal transfer of polycyclic aromatic hydrocarbons in an endangered elasmobranch, the Brazilian guitarfish". *Chemosphere*, 263, 128275. https://doi.org/10.1016/ j.chemosphere.2020.128275

Mastrangelo, G., Fadda, E. & Marzia, V. (1996). "Polycyclic aromatic hydrocarbons and cancer in man". *Environmental Health Perspectives*, 104 (11), 1166–1170. https://doi.org/10.1289/ehp.961041166.

Mazurová, E., Hilscherová, K., Jálová, V., Köhler, H. R., Triebskorn, R., Giesy, J. P. & Bláha, L. (2008). "Endocrine effects of contaminated sediments on the freshwater snail *Potamopyrgus antipodarum in vivo* and in the cell bioassays *in vitro*". *Aquatic toxicology*, 89(3), 172–179. https://doi.org/10.1016/ j.aquatox.2008.06.013.

Meador, J. P., Stein, J. E., Reichert, W. L. & Varanasi, U. (1995). "Bioaccumulation of Polycyclic Aromatic Hydrocarbons by Marine Organisms". *Reviews in Environmental*

Contamination and Toxicology, 143, 79–165. https://doi.org/10.1007/978-1-4612-2542-3_4.

Modica, R., Fiume, M., Guaitani, A. & Bartosek, I. (1983). "Comparative kinetics of benz(a) anthracene, chrysene and triphenylene in rats after oral administration: I. Study with single compounds". Toxicology letters, 18(1-2), 103–109. https://doi.org/10.1016/0378-4274(83)90078-4.

Moiseenko, T. I. (2010). "Effect of Toxic Pollution on Fish Populations and Mechanisms for Maintaining Population Size". Russian Journal of Ecology, 41 (3), 237–243.

Moldan, A. & Jackson, L. F. (1987). Oil spill contingency planning in South Africa. In International Oil Spill Conference, 1, 629B–629B. American Petroleum Institute.

Neff, J. M. (1979). 1979. Polycyclic Aromatic Hydrocarbons in the Aquatic Environment: sources, fates and biological effects. Essex: Applied Science Publishers Ltd.

Neves, E. (2002). Degradação de Hidrocarbonetos Aromáticos Policíclicos por bactérias. Universidade Estadual de Campinas (UNICAMP).

Oliva, A. L., La Colla, N. S., Arias, A. H., Blasina, G. E., Cazorla, A. L. & Marcovecchio, J. E. (2017). "Distribution and human health risk assessment of PAHs in four fish species from a SW Atlantic estuary". Environmental Science and Pollution Research, 24 (23), 18979–18990. https://doi.org/ 10.1007/s11356-017-9394-6.

Patandin, S., Lanting, C. I., Mulder, P. G., Boersma, E. R., Sauer, P. J. & Weisglas-Kuperus, N. (1999). "Effects of environmental exposure to polychlorinated biphenyls and dioxins on cognitive abilities in Dutch children at 42 months of age". Journal of Pediatry, 134, 33-41. https://doi.org/10.1016/s0022-3476 (99)70369-0.

Payne, J. F. & Fancey, L. F. (1989). "Effect of polycyclic aromatic hydrocarbons on immune responses in fish: change in melanomacrophage centers in flounder (Pseudopleuronectes americanus) exposed to hydrocarbon-contaminated sediments." Marine Environmental Research, 28(1-4), 431–435. https://doi.org/10.1016/0141-1136(89)90274-2.

Perugini, M., Visciano, P., Giammarino, A., Manera, M., Di Nardo, W. & Amorena, M. (2007). "Polycyclic Aromatic Hydrocarbons in marine organisms from the Adriatic sea, Italy". Chemosphere, 66(10), 1904–1910. https://doi.org/10.1016/j.chemosphere.2006.07.079.

Phillips, D. H. (1999). "Polycyclic aromatic hydrocarbons in the diet". Mutation Research, 443, 139–147. https://doi.org/10.1016/S1383-5742(99)00016-2.

Ramesh, A., Archibong, A. E., Hood, D. B., Guo, Z. & Loganathan, B. G. (2011). "Global Environmental Distribution and Human Health Effects of Polycyclic Aromatic Hydrocarbons". In: Global Contamination Trends of Persistent Organic Contaminants, edited by Loganathan, B. G.; Lam, P. K. Boca Raton: CRC Press. https://doi.org/10.1201/b11098.

Rigby, C. L., Barreto, R., Carlson, J., Fernando, D., Fordham, S., Francis, M. P., Herman, K., Jabado, R. W., Liu, K. M., Lowe, C. G, Marshall, A., Pacoureau, N., Romanov, E., Sherley, R. B. & Winker, H. (2019). Carcharodon carcharias. The IUCN Red List of Threatened Species 2019: e.T3855A2878674. https://dx.doi.org/10.2305/IUCN.UK.2019-3.RLTS. T3855A2878674.en.

Russel, R. W., Gobas, F. A. P. & Haffner, G. D. (1999). "Maternal transfer and in ovo exposure of organochlorines in oviparous organisms: a model and field verification".

Environmental Science & Technology., 3,3, 416–420. https://doi.org/10.1021/es9800 737.

Salvo, A., et al. (2019). "Accumulation of PCBs, PAHs, plasticizers and inorganic elements in *Hexanchus griseus* from the strait of Messina (Central Mediterranean sea)". *Natural Product Research*, *34* (1), 172-176. https://doi.org/10.1080/14786419.2019.1601197.

SEEDUC. (2019). "Resolução SEEDUC No 5729 de 20 de Março de 2019." ["SEEDUC Resolution No. 5729 of March 20, 2019."] 192. *Diário Oficial Do Estado Do Rio de Janeiro.*

Sun, R., et al. (2019). "Polycyclic aromatic hydrocarbons in marine organisms from Mischief Reef in the South China sea: Implications for sources and human exposure". *Marine Pollution Bulletin*, *149*, 110623.

USEPA. (2000). *Guidance for assessing chemical contaminant data for usein fish advisories. Risk assessment and fish consumption limits*. Office of Water, Washington DC.

Van der Oost, R., Beyer, J. & Vermeulen, N. P. E. (2003). "Fish bioaccumulation and biomarkers in environmental risk assessment: a review". *Environmental Toxicology and Pharmacology*, *13*, 57–149.

VARANASI, U. (1989). *Metabolism of polycyclic aromatic hydrocarbons in the aquatic environment*, Boca Raton: CRC Press, Inc.

Walker, C. H. & Savva, D. (2001). "Biochemical responses of crabs (*Carcinus* spp.) to Polycyclic Aromatic Hydrocarbons (PAHs) as the basis for new biomarker assays". In *Biomarkers in Marine Organisms* (pp. 409-429). Elsevier Science.

Walker, C. J. (2011). *Assessing the Effects of Pollutant Exposure on Sharks: A Biomarker Approach*. Graduate Theses and dissertations, University of North Florida (UNF). Available at https://core.ac.uk/download/pdf/71998765.pdf.

Watson, G. M., Andersen, O. K., Depledge, M. H. & Galloway, T. S. (2004). "Detecting a field gradient of PAH exposure in decapod crustacea using a novel urinary biomarker". *Marine Environmental Research*, *58* (2-5), 257–261.

Whalen, J. (2017). *A multibiomarker analysis of pollutant effects on Atlantic Stingray populations in Florida's St. Johns River*. Master's dissertation, University of North Florida (UNF). Available at https://digitalcommons.unf. edu/cgi/viewcontent.cgi?article=1787&context=etd.

Whyte, J. J., Jung, R. E., Schmitt, C. J. & Tillitt, D. E. (2001). "Ethoxyresorufin-O-deethylase (EROD) Activity in Fish as a Biomarker of Chemical Exposure". *Critical Reviews in Toxicology*, *30* (4), 347–570. https://doi.org/ 10.1080/10408440091159 239.

Zelinkova, Z. & Wenzl, T. (2015). "The occurrence of 16 EPA PAHs in food - A review". *Polycyclic Aromatic Compounds*, *35* (2-4), 248–284. https://doi.org/ 10.1080/ 10406638.2014.918550.

Chapter 3

Phytoremediation of Polycyclic Aromatic Hydrocarbons: From Agricultural Soils to Freshwater Resources

Taylan Kösesakal[*]
Department of Botany, Istanbul University, Istanbul, Turkey

Abstract

Polycyclic aromatic hydrocarbons (PAHs) are hazardous organic compounds with fused aromatic rings originating from incomplete combustion or pyrolysis of fossil fuels. The significant increase in organic pollutants such as petroleum hydrocarbons, PAHs, organic solvents and pesticides worldwide in the last few decades has brought about the necessity to develop new techniques for their removal from the natural environment. Bioremediation, or enhanced biodegradation, is an accepted approach to clean matrices such as soil, sediment, surface and groundwater contaminated with polycyclic aromatic hydrocarbons. A variation of bioremediation is phytoremediation, which is defined as the use of higher plants to remove pollution. PAHs can be adsorbed, accumulated, transported, volatilized, or biodegraded in a non-phytotoxic form by plants. Additionally, plants can degrade organic pollutants by stimulating the microbial community in the rhizosphere. PAHs are known as anthropogenic pollutants harmful to plants, animals and humans. Polycyclic aromatic hydrocarbons threaten human health, agricultural productivity and the environment due to their toxic, mutagenic, carcinogenic and/or persistent properties. The phytoremediation of agricultural soils and freshwater areas contaminated with polycyclic aromatic hydrocarbons is a promising environmentally

---

[*] Corresponding Author's Email: taylank@istanbul.edu.tr.

In: Polycyclic Aromatic Hydrocarbons
Editor: Warren L. Gregoire
ISBN: 978-1-68507-626-9
© 2022 Nova Science Publishers, Inc.

friendly approach. This chapter focused on the potential effects of plants on the phytoremediation of agricultural soils and freshwater resources contaminated with polycyclic aromatic hydrocarbons. The removal of different PAH compounds through different plant species is discussed in plants at the physiological, biochemical and molecular levels and based on phytoremediation mechanisms.

Keywords: polycyclic aromatic hydrocarbons, environmental pollution, phytoremediation, plant stress, plant physiology

Introduction

Industrial development, increasing world population and urban-based environmental pollution threaten natural life, especially since the second half of the twentieth century. In addition, anthropogenic chemicals that the industrialized global economy has gradually released to the environment in the last century have caused pollution in many parts of the world. Pollution occurs because of improper chemical production (oil spills during drilling), transportation (oil spills from tanks or pipelines), storage (chemical spills in storage tanks), usage (such as pesticides and fertilizers in agriculture), or disposal processes (such as explosives in military areas) (Truu et al. 2015). Toxic, mutagenic, carcinogenic and/or persistent pollutants have adverse effects on human health, agricultural productivity and the environment. Examples of these pollutants are total petroleum hydrocarbons (TPH), polycyclic aromatic hydrocarbons (PAH), halogenated hydrocarbons, pesticides, solvents and metals (Saraswat et al. 2021). Polycyclic aromatic hydrocarbons are the most important group of organic pollutants found in terrestrial (Gan, Lau, and Ng 2009) and aquatic environments (Zezulka et al. 2013; Mesa-Marin et al. 2019). Polycyclic aromatic hydrocarbons are known as harmful anthropogenic pollutants for plants, animals and humans (Tomar and Jajoo 2013). PAHs are chemical groups formed during the incomplete combustion of organic molecules and the pyrolysis process of complex series of chemical reactions (Wang et al. 2007; Zhang and Chen 2017; Patel et al. 2020). Most of the PAHs emitted into the environment are due to incomplete combustion of fossil fuels and oil tanker accidents. Due to their mutagenic and carcinogenic properties, PAHs have been included in the list of important pollutants by various environmental and health institutions (Rey-Salgueiro et al. 2008). The United States Environmental Protection Agency (USEPA) declared 16 PAHs (acenaphthene, acenaphthylene, anthracene, benzo(a) anthracene, benzo(a)pyrene, benzo(b)fluoranthene, benzo(g,h,i)perylene, benzo(k)fluoranthene, chrysene, dibenzo(a,h)anthracene, fluoranthene, fluorine, indeno(1,2,3-c,d)pyrene, naphthalene, phenanthrene, pyrene) as priority pollutants in 1983 based on the highest concentration, greater exposure, persistent nature and presence of toxicity (Zheng et al. 2018; Mojiri et al. 2019; Patel et al. 2020). PAHs are ecotoxicologically important because of their persistence in the environment, their tendency to bioaccumulate (Schrenk and Steinberg 1998), their high toxicity, and their mutagenic and carcinogenic potential (Lin et al. 2001). In addition, PAHs have been identified as priority pollutants in environmental pollution monitoring (Srogi 2007).

The increase in the sources of organic and inorganic environmental pollution worldwide has brought with it the necessity of developing techniques to remove them from the natural environment (Samanta, Singh, and Jain 2002). Phytoremediation is an effective method for most organic and inorganic pollutants and includes the following steps: uptake, translocation, transformation, compartmentation, sequestration and mineralization (Pilon-Smits 2005; Yan et al. 2020). The factors affecting the uptake, distribution and transformation of organic compounds by the plant are mostly the physical and chemical properties of the compound (solubility in water, molecular weight, the octanol-water distribution coefficient), as well as environmental conditions (such as temperature, pH, organic matter and soil moisture content) and related to plant properties (such as root system, enzymes) (Susarla, Medina, and McCutcheon 2002; Suresh and Ravishankar 2004). This review contributes to the effects of plants on the phytoremediation of polycyclic aromatic hydrocarbons. In addition, the physiological, biochemical and molecular effects of polycyclic aromatic hydrocarbons on plants were also evaluated.

Effects of Polycyclic Aromatic Hydrocarbons on Plant Growth and Development

Abiotic stress is defined as environmental conditions that reduce plant growth and productivity below the optimum level. Abiotic stress conditions such as organic and inorganic pollution, low temperature, drought, nutrient deficiency, high light and high salinity affect plant growth and productivity. While the productivity of crops and forage crops decreases due to increasing environmental pollution and deteriorating natural balance, our world is faced with a rapidly increasing food requirement with the increasing population (Miao et al. 2015). Therefore, understanding the mechanisms underlying the responses to abiotic stresses is crucial for the maintenance of the agricultural sector as well as the cultivation of stress-tolerant plants (Mahajan and Tuteja 2005). Plants have developed various strategies to cope with abiotic stresses. These include morphological, physiological, biochemical and molecular responses (Naya et al. 2007; Song et al. 2012). All plant growth and development are determined by the interactions between the genome and habitat. When plants are exposed to abiotic stresses, many stress-induced genes contribute to physiological and biochemical responses; stress tolerance, regulation of transcription or signal transduction are activated, and many proteins are produced (Zhuang et al. 2014). Plant phenology is affected by internal or environmental factors (Lee, Lin, and Chiou 2009). Stress conditions can cause the plant to give different physiological responses in different growth periods (Gratani, Crescente, and Rossi 1998). The electron transfer chain is an important area where reactive oxygen species (ROS; $O_2^{\circ-}$, H_2O_2, $\cdot OH$, $^1O_2$) are produced under normal and stress conditions. However, under abiotic stress conditions, ROS generation uses the same electron transfer system, which causes damage to biomolecules (Prasad et al. 2016). Thus, plants have developed an antioxidant system to protect the cellular system from damage caused by excessive ROS production (Kirchsteiger et al. 2009). Enzymes play a crucial role in reducing the effect of oxidative stress by reducing the level of ROS. Enzymes such as superoxide dismutase (SOD), catalase (CAT), peroxidase (POD), ascorbate peroxidase (APX) and glutathione reductase (GR) are vital in combating oxidative stress (Elavarthi and Martin 2010). The plant defence mechanism against adverse growth conditions such as drought, nutrient deficiency, light, high temperature, salinity, organic and inorganic pollutants is based on the activation of secondary metabolism, which results in the accumulation of various antioxidants to maintain the balance between antioxidant/oxidant species (Herms and Mattson 1992). Antioxidant compounds are represented by ascorbic acid, tocopherols, carotenoids, and some phenolic compounds, which can be enzymatic and non-enzymatic. These compounds directly affect plants' chemical composition and health (McKersie and Leshem 1994; Berger 2005). The content and distribution of common plant pigments such as chlorophyll, carotenoids and anthocyanins determine both the colour and appearance of the plant (Abbott 1999). Pigment synthesis in plants may be the result of biotic or abiotic stress conditions, senescence, or ecological adaptation to changing environmental conditions (Gould et al. 1995). Therefore, chlorophylls, carotenoids and flavonoids maintain a balanced physiological state in plant tissues (Stintzing and Carle 2004). Furthermore, plants produce and accumulate various metabolic products, including phenolics and flavonoids, as secondary plant metabolites under stress conditions (Grace and Logan 2000). These compounds play an important role in processes such as protection, repair and degradation under stress conditions caused by toxic chemicals (Rice-Evans, Miller, and Paganga 1997). Anthocyanins are partially responsible for the colour of plant tissue and are transported to the vacuole after being produced in the cytoplasm (Shirley 1996). Anthocyanin synthesis is promoted by UV-B, nutrient deficiencies, low temperatures, and heavy metal stresses (Warren et al. 2003; Pinto et al. 1999; Rabino and Mancinelli 1986; Dai, Dong, and Ma 2012; Kosesakal 2014).

Contamination of organic pollutants, including PAHs, not only affects water and soil quality but also has adverse effects on organisms or biotic associations. The presence of these pollutants in the living environments of plants creates negative effects for plant growth and development (Zezulka et al. 2013). As with other toxic organic compounds, PAHs have been proven to affect the biochemical and physiological processes in plants (Liu et al. 2015; Fahid et al. 2020). The responses of plants to PAH compounds vary according to PAH species and concentrations. PAHs affect not only energy metabolism processes but also mechanisms associated with plant growth and development. In higher plants, PAHs can cause decreased biomass accumulation, chlorosis, and inhibition of photosynthesis (Oguntimehin, Nakatani, and Sakugawa 2008). As with other stress factors, the first effects of PAHs are on chloroplasts (Ahammed et al. 2012). PAHs preferentially accumulate in chloroplast thylakoids and microsomes, where they affect primary photochemical processes (Huang et al. 1997). PAHs application accelerates the senescence and death of roots and stimulates the formation of new roots (Li et al. 2021). In addition, PAHs decrease the photosynthetic pigment content by direct interaction with pigment molecules or inhibition of their synthesis (Kummerova et al. 2006; Oguntimehin, Eissa, and Sakugawa 2010). These pollutants may have effects such as delaying seed germination, reducing plant growth, photosynthesis rate and biomass, or may cause complete death of the plant (Salehi-Lisar, Deljoo, and Tejada Moral 2015; Afegbua and Batty 2018; Kreslavski et al. 2017).

Phytoremediation of Polycyclic Aromatic Hydrocarbons

It is of great importance to evaluate the potential carcinogenic risks posed by PAHs in soil, sediment and freshwaters and to remove the pollutants in these areas. Remediation of aromatic compounds and other organic pollutants is a complex process. This process depends on many factors such as the plant, the rhizospheric microorganisms, and the bioavailability of the pollutant by the plant and microorganisms. As the molecular weight of the PAH compound and the benzene ring it contains increase, its degradation slows down. Surface waters, soil, and sediment can be exposed to PAH pollution through oil spills, wastewater, domestic and industrial waste. In the aquatic environment, lipophilic PAHs can exist in both dissolved form and adsorbed on solid particles. Half-lives of PAHs can be from one week to two months in aquatic environments and from six months to six years in sediment (Kalf, Crommentuijn, and van de Plassche 1997). Although aromatic hydrocarbons are more biodegradation-resistant, some low molecular weight aromatic compounds such as naphthalene can oxidize earlier than most saturated hydrocarbons (Foght and Westlake 1987; Zhu et al. 2001). Monoaromatic hydrocarbons are toxic to some microorganisms due to their solvent effects on cell membranes. However, under aerobic conditions, they can readily biodegrade at low concentrations. 2 – 4 ring PAHs are less toxic and their biodegradation rate decreases as their structure become more complex. The biodegradation of PAH compounds with 5 or more rings can occur with the co-metabolism of microorganisms (Prince 1993; Cerniglia 1993; Zhu et al. 2001).

The seedling growth and growth rates of plants grown in contaminated areas can be used to determine their phytoremediation potential. The availability of the plant in phytoremediation of these areas can be estimated by considering germination rates, plant length, root and stem biomass. Root biomass is the most important of these parameters in the evaluation of plant phytoremediation potential in PAH-contaminated areas (Issoufi, Rhykerd, and Smiciklas 2006). Effective degradation of pollutants by phytoremediation depends on an extensive root system. Roots increase the efficiency of phytoremediation by sequestering pollutants, providing soil aeration and restricting the movement of pollutants in the soil. The most important contribution of roots in the phytoremediation of organic pollutants is undoubted that they increase the activity of microorganisms (Jamie, Kevin, and Robert 2016). Because with increasing root biomass, the rhizosphere effect also increases. The population of microorganisms in the rhizosphere is

10 – 100 times higher than in areas without rhizosphere (Issoufi, Rhykerd, and Smiciklas 2006). Ling and Gao (2004) reported that the loss of phenanthrene and pyrene in *Amaranthus tricolor* cultivated soil was 87.85% - 94.03% and 46.89% – 76.57%, but this loss was only between 2.55% - 13.66% and 11.12% - 56.55% in unplanted soil. Wei and Pan (2010) stated that *Medicago sativa*, *Brassica campestris* and *Trifolium repens* plants promote phenanthrene and pyrene degradation. Shahsavari et al. (2015) reported that wheat was 98% - 100% and 65% - 70% effective in the biodegradation of phenanthrene and pyrene, respectively. They also noted that the wheat rhizosphere increased the total microbial abundance including PAH-degrading organisms, and these increased activities resulted in enhanced degradation of phenanthrene and pyrene. Jajoo et al. (2014) investigated the inhibitory effects of three PAH compounds (naphthalene, anthracene and pyrene) on the photosynthetic process in *Arabidopsis thaliana* grown in soil. Their results showed that all PAH species inhibited PS II photochemistry, protein composition and photosynthetic performance and reduced pigment concentration. Jeelani et al. (2017) investigated the independent and interactive effects of cadmium and PAHs (phenanthrene and pyrene) on the growth of the wetland plant *Acorus calamus*. Furthermore, they studied the plant's ability to uptake, accumulate, and remove pollutants from soils. The correlation analysis of their results indicated a positive relationship between residual concentrations of phenanthrene and pyrene, whereas enzyme activities (dehydrogenase and polyphenol oxidase) were negatively correlated with each other. Kosnar, Mercl, and Tlustos (2018) reported that the presence of *Zea mays* significantly increased the PAH removal from soil and its above-ground biomass did not represent any environmental risk. Thus, they stated that maize plants could be used to enhance bioremediation of PAHs in agricultural soils and *in situ* PAH phytoremediation. A three-year phytoremediation experiment with willow showed that the total removal of PAHs from the soil was 50.9% (Kosnar, Mercl, and Tlustos 2020). Plant and microbial assisted phytoremediation have some potential as an effective and inexpensive means to clean up PAH-polluted soils. It is stated that the symbiotic association between *Medicago sativa* L. and *Rhizobium meliloti* can stimulate the rhizosphere microflora to degrade PAHs, and its application may be a promising bioremediation strategy for PAH-contaminated soils (Teng et al. 2011). Xiao et al. (2015) conducted pot-culture experiments to assess the phytoremediation potential of five ornamental plant species on PAH contaminated soils. Their results indicated that Fire Phoenix and *Medicago sativa* L effectively reduced the PAH content

in contaminated soil, and thus they stated that these two plants could be used for the phytoremediation of PAH-contaminated soils.

Aquatic plants (from algae to macrophytic species) are crucial organisms for sediment stabilization, water quality control, nutrient cycling and oxygen production, as well as the protection and habitat provision of aquatic organisms. Additionally, the stability and degrading activities of aquatic plants and microorganisms against pollutants provide biotechnological solutions for the phytoremediation of aquatic fields (Tumaikina, Turkovskaya, and Ignatov 2008). Maillacheruvu and Safaai (2002) stated that in hydroponic culture, *Sagittaria* removed 40% of naphthalene in the presence of light. Wand et al. (2002) reported that naphthalene was effectively eliminated from the environment by *Carex gracilis*. Kosesakal et al. (2016) investigated the phytoremediation capacity of *Azolla filiculoides* Lam. for the water resources contaminated with petroleum hydrocarbons under laboratory conditions. Compared to control samples, the biodegradation rate of phenanthrene at oil concentrations of 0.05% - 0.2% was 81% - 77%. Thus, our results suggested that *A. filiculoides* could be a promising candidate to be used for the phytoremediation of low crude oil contaminated freshwater resources. Kosesakal et al. (2015) stated that the presence of crude oil up to 0.5% v/v reduced growth by as much as 50% relative to the control *Lemna minor* plants. At 0.5% and 1% oil applications, we found that the plant samples contained 5-ring polycyclic aromatic hydrocarbons; their densities were approximately two times lower than the density of the unplanted control samples. Thus, we concluded that the biodegradation potential and the phytoremediation capacity of *L. minor* strongly depend on the concentration of crude oil contaminants. Ertekin et al. (2015) stated that crude oil negatively affected the growth of *Landoltia punctata*, a freshwater plant. Verane et al. (2020) reported that *Rhizophora mangle* promoted the degradation of 16 PAHs in mangrove sediments and showed better phytoremediation efficiency (60.76%) compared to natural attenuation (49.57%). Liu et al. (2014) investigated the effect of plant density of *Vallisneria spiralis* on phytoremediation of polycyclic aromatic hydrocarbons contaminated sediments. They stated that accelerated removal of PAHs (phenanthrene and pyrene) in sediments in the presence of *V. spiralis*. Thus, they indicated that microbial degradation played a major role in the *V. spiralis*-promoted remediation, and it is concluded that the enhanced dissipation of PAHs is mainly related to the oxygen released by roots.

The extent to which PAH exposure triggers stress signalling pathways common to other abiotic or biotic stresses in plants and whether PAH stress-specific signalling components exist is unclear (Singha and Pandey 2017).

Weisman, Alkio, and Colón-Carmona (2010) reported that phenanthrene treatment caused the downregulation of genes involved in photosynthesis and protein biosynthesis in *Arabidopsis thaliana*. Furthermore, their results suggested that the complex physiological PAH stress symptoms likely involve multiple hormone pathways, including salicylic acid, ethylene, jasmonic acid, and abscisic acid (ABA). Peng et al. (2014) reported that phenanthrene was removed from the soil and aquatic environment at a higher rate than wild species in transgenic *Arabidopsis* and rice plants transferred to the naphthalene dioxygenase system. Liu et al. (2015) indicated that the nucleoside diphosphate kinase 3 (NDPK3) is a positive regulator in the *A. thaliana* under the phenanthrene stress at the proteomic level. Hernandez-Vega et al. (2017) reported that AT5G05600, a putative flavonol synthase, is part of the phenanthrene-induced stress response and is probably involved in the first detoxification phase for PAHs in mutant *Arabidopsis thaliana*. Wild et al. (2005) reported that apoplastic transport seems to be the dominant pathway for root transport in a study with two-photon excitation microscopy (TPEM) investigating the uptake and transport of corn and wheat roots grown in soil contaminated with phenanthrene and anthracene compounds. Simultaneously, anthracene degradation was directly observed in the cell walls of metabolically active mature root cortex cells of both plant species. Alkio et al. (2005) reported that the gene expression of the cell wall-loosening protein expansin was repressed, whereas the gene expression of the pathogenesis-related protein PR1 was induced in response to phenanthrene exposure. Li et al. (2021) reported that phenanthrene affected the expression of CDK (the coding gene of cyclin-dependent kinase) and CDC2 (a gene regulating cell division cycle), the key genes in the cell cycle of pericycle cells, thereby affecting the occurrence and growth of lateral roots. Furthermore, they stated that microRNAs (miR164) could regulate root growth and adventitious root generation of *Triticum aestivum* L. under phenanthrene exposure by targeting the NAC transcription factor. Guo et al. (2018) investigated the effects of tall fescue on the removal of PAHs and PAH-degrading genes. They found that tall fescue improved PAH removal from the contaminated soil and stimulated the activities of enzymes encoded by the RHDα GN genes of the bacteria in the rhizosphere soil.

Conclusion

In phytoremediation studies to eliminate PAH pollution; plants, plants and microorganisms, additives in the form of plants and fertilizers, multi-faceted processes where plants, microorganisms, and additives are used together give effective results. In areas where plants and/or other treatments are used together in PAH phytoremediation studies, degradation is greater than in unplanted environments. Phytoremediation of water basins can be carried out with the participation of microbial association of various ecological groups, bacterial periphyton which covers solid surfaces, for example, the surface of aquatic plants. Plants respond to pollution by absorbing, accumulating or transforming xenobiotics through their enzymes, apart from supporting the substrate (Tumaikina, Turkovskaya, and Ignatov 2008). Additionally, phenolic and flavonoids secreted by plants can facilitate the transformation of organic pollutants (Ite and Semple 2015). The diversity of substrates that degrade in the presence of secondary plant metabolites can be explained by considering the similar skeletal characteristics of pollutants and secondary plant metabolites (Singer, Crowley, and Thompson 2003). The stability and degrading activities of plants and associated microorganisms against pollutants offer biotechnological solutions for phytoremediation of PAHs-contaminated areas. The use of plants in the removal of PAH pollution leads to more effective degradation than unplanted environments. Complex interactions of plants and plant-associated microorganisms play a crucial role in the biodegradation of high molecular weight PAHs and other complex molecules. Therefore, for a sound evaluation of phytoremediation, it is crucial to understand the complex interaction involving the ecological, physiological and molecular mechanisms of the plant and associated microorganisms that affect the biodegradation of pollutants. Alternatively, although the focus of phytoremediation is to reduce environmental concentrations, plant accumulation of PAHs is also important due to potential food chain issues (Schwab and Dermody 2021). In this context, it can be stated that the data obtained from the following multidisciplinary studies will provide important contributions to reveal the effects of PAH compounds on plants and microorganisms and to evaluate the phytoremediation of plants on PAH compounds in detail; *i*) detailed biochemical and physiological analysis of antioxidant enzyme systems and photosynthetic process *ii*) investigation of the stress response at the molecular level *iii*) examining the molecular mechanism of PAH degradation *iv*) detection of plant and rhizospheric bacteria and their effects on the degradation of PAH compounds.

Acknowledgment

I would like to thank Istanbul University Scientific Research Projects Coordination Unit for supporting my projects in this field.

References

Abbott, Judith A. 1999. "Quality measurement of fruits and vegetables." *Postharvest Biology and Technology* 15 (3):207-225. doi: https://doi. org/10.1016/S0925-5214(98)00086-6.

Afegbua, S. L., and L. C. Batty. 2018. "Effect of single and mixed polycyclic aromatic hydrocarbon contamination on plant biomass yield and PAH dissipation during phytoremediation." *Environ Sci Pollut Res Int* 25 (19):18596-18603. doi: 10.1007/s11356-018-1987-1.

Ahammed, G. J., M. M. Wang, Y. H. Zhou, X. J. Xia, W. H. Mao, K. Shi, and J. Q. Yu. 2012. "The growth, photosynthesis and antioxidant defense responses of five vegetable crops to phenanthrene stress." *Ecotoxicol Environ Saf* 80:132-9. doi: 10.1016/j.ecoenv.2012.02.015.

Alkio, M., T. M. Tabuchi, X. Wang, and A. Colon-Carmona. 2005. "Stress responses to polycyclic aromatic hydrocarbons in *Arabidopsis* include growth inhibition and hypersensitive response-like symptoms." *J Exp Bot* 56 (421):2983-94. doi: 10.1093/jxb/ eri295.

Berger, M. M. 2005. "Can oxidative damage be treated nutritionally?" *Clin Nutr* 24 (2):172-83. doi: 10.1016/j.clnu.2004.10.003.

Cerniglia, Carl E. 1993. "Biodegradation of polycyclic aromatic hydrocarbons." *Current Opinion in Biotechnology* 4 (3):331-338. doi: https://doi.org/ 10.1016/0958-1669(93)90104-5.

Dai, L. P., X. J. Dong, and H. H. Ma. 2012. "Molecular mechanism for cadmium-induced anthocyanin accumulation in *Azolla imbricata*." *Chemosphere* 87 (4):319-25. doi: 10.1016/j.chemosphere.2011. 12.005.

Elavarthi, Sathya, and Bjorn Martin. 2010. "Spectrophotometric Assays for Antioxidant Enzymes in Plants." In *Plant Stress Tolerance: Methods and Protocols*, edited by Ramanjulu Sunkar, 273-280. Totowa, NJ: Humana Press.

Ertekin, O., T. Kosesakal, V.S. Unlu, S. Dagli, V. Pelitli, H. Uzyol, Y. Tuna, O. Kulen, B. Yuksel, S. Onarici, B. C. Keskin, and A. Memon. 2015. "Phytoremediation potential of *Landoltia punctata* on petroleum hydrocarbons." *Turkish Journal of Botany* 39 (1):23-29. doi: 10.3906/bot-1403-42.

Fahid, M., M. Arslan, G. Shabir, S. Younus, T. Yasmeen, M. Rizwan, K. Siddique, S. R. Ahmad, R. Tahseen, S. Iqbal, S. Ali, and M. Afzal. 2020. "*Phragmites australis* in combination with hydrocarbons degrading bacteria is a suitable option for remediation of diesel-contaminated water in floating wetlands." *Chemosphere* 240:124890. doi: 10.1016/j.chemosphere.2019.124890.

Foght, J. M., and D. W. S. Westlake. 1987. "Biodegradation of Hydrocarbons in Freshwater." In *Oil in Freshwater: Chemistry, Biology, Countermeasure Technology*, edited by John H. Vandermeulen and Steve E. Hrudey, 217-230. Pergamon.
Gan, S., E. V. Lau, and H. K. Ng. 2009. "Remediation of soils contaminated with polycyclic aromatic hydrocarbons (PAHs)." *J Hazard Mater* 172 (2-3):532-49. doi: 10.1016/j.jhazmat.2009.07.118.
Gould, Kevin S., David N. Kuhn, David W. Lee, and Steven F. Oberbauer. 1995. "Why leaves are sometimes red." *Nature* 378 (6554):241-242. doi: 10.1038/378241b0.
Grace, S. C., and B. A. Logan. 2000. "Energy dissipation and radical scavenging by the plant phenylpropanoid pathway." *Philos Trans R Soc Lond B Biol Sci* 355 (1402):1499-510. doi: 10.1098/rstb. 2000.0710.
Gratani, L., M. F. Crescente, and G. Rossi. 1998. "Photosynthetic Performance and Water Use Efficiency of the Fern *Cheilanthes persica*." *Photosynthetica* 35 (4):507-516. doi: 10.1023/A:10069 70705546.
Guo, M., Z. Gong, R. Miao, C. Jia, J. Rookes, D. Cahill, and J. Zhuang. 2018. "Enhanced polycyclic aromatic hydrocarbons degradation in rhizosphere soil planted with tall fescue: Bacterial community and functional gene expression mechanisms." *Chemosphere* 212:15-23. doi: 10.1016/j.chemosphere.2018. 08.057.
Herms, Daniel A., and William J. Mattson. 1992. "The Dilemma of Plants: To Grow or Defend." *The Quarterly Review of Biology* 67 (3):283-335.
Hernandez-Vega, J. C., B. Cady, G. Kayanja, A. Mauriello, N. Cervantes, A. Gillespie, L. Lavia, J. Trujillo, M. Alkio, and A. Colon-Carmona. 2017. "Detoxification of polycyclic aromatic hydrocarbons (PAHs) in *Arabidopsis thaliana* involves a putative flavonol synthase." *J Hazard Mater* 321:268-280. doi: 10.1016/j.jhazmat.2016. 08.058.
Huang, Xiao-Dong, Brendan J. McConkey, T. Sudhakar Babu, and Bruce M. Greenberg. 1997. "Mechanisms of photoinduced toxicity of photomodified anthracene to plants: Inhibition of photosynthesis in the aquatic higher plant *Lemna gibba* (duckweed)." *Environmental Toxicology and Chemistry* 16 (8):1707-1715. doi: https://doi.org/10. 1002/etc.5620160819.
Issoufi, I., R. L. Rhykerd, and K. D. Smiciklas. 2006. "Seedling Growth of Agronomic Crops in Crude Oil Contaminated Soil." *Journal of Agronomy and Crop Science* 192 (4):310-317. doi: https://doi.org/ 10.1111/j.1439-037X.2006.00212.x.
Ite, Aniefiok E., and Kirk T. Semple. 2015. "The Effect of Flavonoids on the Microbial Mineralisation of Polycyclic Aromatic Hydrocarbons in Soil." *International Journal of Environmental Bioremediation & Biodegradation* 3 (3):66-78. doi: 10.12691/ijebb-3-3-1.
Jajoo, A., N. R. Mekala, R. S. Tomar, M. Grieco, M. Tikkanen, and E. M. Aro. 2014. "Inhibitory effects of polycyclic aromatic hydrocarbons (PAHs) on photosynthetic performance are not related to their aromaticity." *J Photochem Photobiol B* 137:151-5. doi: 10.1016/j. jphotobiol.2014.03.011.
Jamie, D. Spiares, E. Kenworthy Kevin, and L. Rhykerd Robert. 2016. "Root and Shoot Biomass of Plants Seeded in Crude Oil Contaminated Soil." *Texas Journal of Agriculture and Natural Resources* 14 (0):117-124.

Jeelani, N., W. Yang, L. Xu, Y. Qiao, S. An, and X. Leng. 2017. "Phyto-remediation potential of *Acorus calamus* in soils co-contaminated with cadmium and polycyclic aromatic hydrocarbons." *Sci Rep* 7 (1):8028. doi: 10.1038/s41598-017-07831-3.

Kalf, Dennis F., Trudie Crommentuijn, and Erik J. van de Plassche. 1997. "Environmental Quality Objectives for 10 Polycyclic Aromatic Hydrocarbons (PAHs)." *Ecotoxicology and Environmental Safety* 36 (1):89-97. doi: https://doi.org/10.1006/eesa.1996.1495.

Kirchsteiger, K., P. Pulido, M. Gonzalez, and F. J. Cejudo. 2009. "NADPH Thioredoxin reductase C controls the redox status of chloroplast 2-Cys peroxiredoxins in *Arabidopsis thaliana*." *Mol Plant* 2 (2):298-307. doi: 10.1093/mp/ssn082.

Kosesakal, T., M. Unal, O. Kulen, A. Memon, and B. Yuksel. 2016. "Phytoremediation of petroleum hydrocarbons by using a freshwater fern species *Azolla filiculoides* Lam." *Int J Phytoremediation* 18 (5):467-76. doi: 10.1080/15226514.2015.1115958.

Kosesakal, T. 2014. "Effects of Seasonal Changes on Pigment Composition of *Azolla filiculoides* Lam." *American Fern Journal* 104 (2):58-66.

Kosesakal, T., V.S. Unlu, O. Kulen, A. Memon, and B. Yuksel. 2015. "Evaluation of the phytoremediation capacity of *Lemna minor* L. in crude oil spiked cultures." *Turkish Journal of Biology* 39 (3):479-484. doi: 10.3906/ biy-1406-85.

Kosnar, Z., F. Mercl, and P. Tlustos. 2018. "Ability of natural attenuation and phytoremediation using maize (*Zea mays* L.) to decrease soil contents of polycyclic aromatic hydrocarbons (PAHs) derived from biomass fly ash in comparison with PAHs-spiked soil." *Ecotoxicol Environ Saf* 153:16-22. doi: 10.1016/j.ecoenv.2018.01.049.

Kosnar, Z., F. Mercl, and P. Tlustos. 2020. "Long-term willows phytoremediation treatment of soil contaminated by fly ash polycyclic aromatic hydrocarbons from straw combustion." *Environ Pollut* 264:114787. doi: 10.1016/j.envpol.2020.114787.

Kreslavski, V. D., M. Brestic, S. K. Zharmukhamedov, V. Y. Lyubimov, A. V. Lankin, A. Jajoo, and S. I. Allakhverdiev. 2017. "Mechanisms of inhibitory effects of polycyclic aromatic hydrocarbons in photosynthetic primary processes in pea leaves and thylakoid preparations." *Plant Biol (Stuttg)* 19 (5):683-688. doi: 10.1111/ plb.12598.

Kummerova, M., J. Krulova, S. Zezulka, and J. Triska. 2006. "Evaluation of fluoranthene phytotoxicity in pea plants by Hill reaction and chlorophyll fluorescence." *Chemosphere* 65 (3):489-96. doi: 10.1016/j.chemosphere. 2006.01.052.

Lee, P. H., T. T. Lin, and W. L. Chiou. 2009. "Phenology of 16 species of ferns in a subtropical forest of northeastern Taiwan." *J Plant Res* 122 (1):61-7. doi: 10.1007/s10265-008-0191-7.

Li, J., H. Zhang, J. Zhu, Y. Shen, N. Zeng, S. Liu, H. Wang, J. Wang, and X. Zhan. 2021. "Role of miR164 in the growth of wheat new adventitious roots exposed to phenanthrene." *Environ Pollut* 284:117204. doi: 10.1016/ j.envpol.2021.117204.

Lin, Chin H., Xuanwei Huang, Alexander Kolbanovskii, Brian E. Hingerty, Shantu Amin, Suse Broyde, Nicholas E. Geacintov, and Dinshaw J. Patel. 2001. "Molecular topology of polycyclic aromatic carcinogens determines DNA adduct conformation: a link to tumorigenic activity1 1Edited by M. F. Summers." *Journal of Molecular Biology* 306 (5):1059-1080. doi: https://doi.org/10. 1006/jmbi.2001.4425.

Ling, Wanting, and Yanzheng Gao. 2004. "Promoted dissipation of phenanthrene and pyrene in soils by amaranth (*Amaranthus tricolor* L.)." *Environmental Geology* 46 (5). doi: 10.1007/s00254-004-1028-x.

Liu, H., D. Weisman, L. Tang, L. Tan, W. K. Zhang, Z. H. Wang, Y. H. Huang, W. X. Lin, X. M. Liu, and A. Colon-Carmona. 2015. "Stress signaling in response to polycyclic aromatic hydrocarbon exposure in *Arabidopsis thaliana* involves a nucleoside diphosphate kinase, NDPK-3." *Planta* 241 (1):95-107. doi: 10.1007/s00425-014-2161-8.

Liu, Hongyan, Fanbo Meng, Yindong Tong, and Jie Chi. 2014. "Effect of plant density on phytoremediation of polycyclic aromatic hydrocarbons contaminated sediments with *Vallisneria spiralis*." *Ecological Engineering* 73:380-385. doi: 10.1016/j.ecoleng.2014.09.084.

Mahajan, S., and N. Tuteja. 2005. "Cold, salinity and drought stresses: an overview." *Arch Biochem Biophys* 444 (2):139-58. doi: 10.1016/j.abb.2005.10.018.

Maillacheruvu, K., and S. Safaai. 2002. "Naphthalene removal from aqueous systems by *Sagittarius* sp." *J Environ Sci Health A Tox Hazard Subst Environ Eng* 37 (5):845-61. doi: 10.1081/ese-120003592.

McKersie, Bryan D., and Ya'acov Y. Leshem. 1994. "Chilling stress." In *Stress and Stress Coping in Cultivated Plants*, edited by Bryan D. McKersie and Ya'acov Y. Leshem, 79-103. Dordrecht: Springer Netherlands.

Mesa-Marin, J., J. M. Barcia-Piedras, E. Mateos-Naranjo, L. Cox, M. Real, J. A. Perez-Romero, S. Navarro-Torre, I. D. Rodriguez-Llorente, E. Pajuelo, R. Parra, and S. Redondo-Gomez. 2019. "Soil phenanthrene phytoremediation capacity in bacteria-assisted *Spartina densiflora*." *Ecotoxicol Environ Saf* 182:109382. doi: 10.1016/j.ecoenv.2019.109382.

Miao, Z., W. Xu, D. Li, X. Hu, J. Liu, R. Zhang, Z. Tong, J. Dong, Z. Su, L. Zhang, M. Sun, W. Li, Z. Du, S. Hu, and T. Wang. 2015. "De novo transcriptome analysis of *Medicago falcata* reveals novel insights about the mechanisms underlying abiotic stress-responsive pathway." *BMC Genomics* 16:818. doi: 10.1186/s12864-015-2019-x.

Mojiri, A., J. L. Zhou, A. Ohashi, N. Ozaki, and T. Kindaichi. 2019. "Comprehensive review of polycyclic aromatic hydrocarbons in water sources, their effects and treatments." *Sci Total Environ* 696:133971. doi: 10.1016/j.scitotenv.2019.133971.

Naya, L., R. Ladrera, J. Ramos, E. M. Gonzalez, C. Arrese-Igor, F. R. Minchin, and M. Becana. 2007. "The response of carbon metabolism and antioxidant defenses of alfalfa nodules to drought stress and to the subsequent recovery of plants." *Plant Physiol* 144 (2):1104-14. doi: 10.1104/pp.107.099648.

Oguntimehin, I., F. Eissa, and H. Sakugawa. 2010. "Negative effects of fluoranthene on the ecophysiology of tomato plants (*Lycopersicon esculentum* Mill) Fluoranthene mists negatively affected tomato plants." *Chemosphere* 78 (7):877-84. doi: 10.1016/j.chemosphere.2009.11.030.

Oguntimehin, I., N. Nakatani, and H. Sakugawa. 2008. "Phytotoxicities of fluoranthene and phenanthrene deposited on needle surfaces of the evergreen conifer, Japanese red pine (*Pinus densiflora* Sieb. et Zucc.)." *Environ Pollut* 154 (2):264-71. doi: 10.1016/j.envpol.2007.10.039.

Patel, A. B., S. Shaikh, K. R. Jain, C. Desai, and D. Madamwar. 2020. "Polycyclic Aromatic Hydrocarbons: Sources, Toxicity, and Remediation Approaches." *Front Microbiol* 11:562813. doi: 10.3389/fmicb.2020.562813.

Peng, R. H., X. Y. Fu, W. Zhao, Y. S. Tian, B. Zhu, H. J. Han, J. Xu, and Q. H. Yao. 2014. "Phytoremediation of phenanthrene by transgenic plants transformed with a naphthalene dioxygenase system from Pseudomonas." *Environ Sci Technol* 48 (21):12824-32. doi: 10.1021/es5015357.

Pilon-Smits, E. 2005. "Phytoremediation." *Annu Rev Plant Biol* 56:15-39. doi: 10.1146/annurev.arplant.56.032604.144214.

Pinto, Manuel E., Paula Casati, Tsui-Ping Hsu, Maurice S. B. Ku, and Gerald E. Edwards. 1999. "Effects of UV-B radiation on growth, photosynthesis, UV-B-absorbing compounds and NADP-malic enzyme in bean (*Phaseolus vulgaris* L.) grown under different nitrogen conditions." *Journal of Photochemistry and Photobiology B: Biology* 48 (2):200-209. doi: https://doi.org/10.1016/S1011-1344(99)00031-7.

Prasad, S. M., S. Kumar, P. Parihar, and R. Singh. 2016. "Interactive effects of herbicide and enhanced UV-B on growth, oxidative damage and the ascorbate-glutathione cycle in two *Azolla* species." *Ecotoxicol Environ Saf* 133:341-9. doi: 10.1016/j.ecoenv. 2016. 07.036.

Prince, Roger C. 1993. "Petroleum Spill Bioremediation in Marine Environments." *Critical Reviews in Microbiology* 19 (4):217-240. doi: 10.3109/10408419309113530.

Rabino, I., and A. L. Mancinelli. 1986. "Light, temperature, and anthocyanin production." *Plant physiology* 81 (3):922-924. doi: 10.1104/pp.81.3.922.

Rey-Salgueiro, Ledicia, Elena Martínez-Carballo, Mercedes Sonia García-Falcón, and Jesús Simal-Gándara. 2008. "Effects of a chemical company fire on the occurrence of polycyclic aromatic hydrocarbons in plant foods." *Food Chemistry* 108 (1):347-353. doi: 10.1016/j.foodchem.2007.10.042.

Rice-Evans, Catherine, Nicholas Miller, and George Paganga. 1997. "Antioxidant properties of phenolic compounds." *Trends in Plant Science* 2 (4):152-159. doi: https://doi.org/10.1016/S1360-1385(97)01018-2.

Salehi-Lisar, Seyed Yahya, Somayeh Deljoo, and Manuel Tejada Moral. 2015. "The physiological effect of fluorene on Triticum aestivum, *Medicago sativa*, and *Helianthus annus*." *Cogent Food & Agriculture* 1 (1). doi: 10.1080/23311932. 2015.1020189.

Samanta, Sudip K., Om V. Singh, and Rakesh K. Jain. 2002. "Polycyclic aromatic hydrocarbons: environmental pollution and bioremediation." *Trends in Biotechnology* 20 (6):243-248. doi: https://doi.org/10.1016/S0167-7799 (02)01943-1.

Saraswat, Pooja, Kritika Yadav, Anamika Gupta, Mrinalini Prasad, and Rajiv Ranjan. 2021. "Physiological and molecular basis for remediation of polyaromatic hydrocarbons." In *Handbook of Bioremediation*, 535-550.

Schrenk, C., and C. Steinberg. 1998. "Metabolism of phenanthrene in cell suspension cultures of wheat and soybean as well as in intact plants of the water moss *Fontinalis antipyretica*: A comparative study." *Environmental science and pollution research international* 5 2:83-8.

Schwab, A. P., and C. L. Dermody. 2021. "*Pathways of polycyclic aromatic hydrocarbons assimilation by plants growing in contaminated soils.*" In, 193-250.

Shahsavari, E., E. M. Adetutu, M. Taha, and A. S. Ball. 2015. "Rhizoremediation of phenanthrene and pyrene contaminated soil using wheat." *J Environ Manage* 155:171-6. doi: 10.1016/j.jenvman.2015.03.027.

Shirley, Brenda W. 1996. "Flavonoid biosynthesis: 'new' functions for an 'old' pathway." *Trends in Plant Science* 1 (11):377-382. doi: https://doi.org/ 10.1016/S1360-1385(96)80312-8.

Singer, Andrew C., David E. Crowley, and Ian P. Thompson. 2003. "Secondary plant metabolites in phytoremediation and biotransformation." *Trends in Biotechnology* 21 (3):123-130. doi: 10.1016/s0167-7799(02)00041-0.

Singha, L. Paikhomba, and Piyush Pandey. 2017. "Glutathione and glutathione-S-transferase activity in *Jatropha curcas* in association with pyrene degrader *Pseudomonas aeruginosa* PDB1 in rhizosphere, for alleviation of stress induced by polyaromatic hydrocarbon for effective rhizoremediation." *Ecological Engineering* 102:422-432. doi: 10.1016/j.ecoleng.2017.02.061.

Song, Shiyong, Ying Chen, Mingui Zhao, and Wen-Hao Zhang. 2012. "A novel *Medicago truncatula* HD-Zip gene, MtHB2, is involved in abiotic stress responses." *Environmental and Experimental Botany* 80:1-9. doi: 10.1016/ j.envexpbot. 2012.02.001.

Srogi, K. 2007. "Monitoring of environmental exposure to polycyclic aromatic hydrocarbons: a review." *Environ Chem Lett* 5 (4):169-195. doi: 10.1007/ s10311-007-0095-0.

Stintzing, Florian C., and Reinhold Carle. 2004. "Functional properties of anthocyanins and betalains in plants, food, and in human nutrition." *Trends in Food Science & Technology* 15 (1):19-38. doi: 10.1016/j. tifs.2003.07.004.

Suresh, B., and G. A. Ravishankar. 2004. "Phytoremediation—A Novel and Promising Approach for Environmental Clean-up." *Critical Reviews in Biotechnology* 24 (2-3):97-124. doi: 10.1080/073885 50490493627.

Susarla, Sridhar, Victor F. Medina, and Steven C. McCutcheon. 2002. "Phytoremediation: An ecological solution to organic chemical contamination." *Ecological Engineering* 18 (5):647-658. doi: https://doi.org/ 10.1016/S0925-8574(02)00026-5.

Teng, Y., Y. Shen, Y. Luo, X. Sun, M. Sun, D. Fu, Z. Li, and P. Christie. 2011. "Influence of Rhizobium meliloti on phytoremediation of polycyclic aromatic hydrocarbons by alfalfa in an aged contaminated soil." *J Hazard Mater* 186 (2-3):1271-6. doi: 10.1016/j.jhazmat. 2010.11.126.

Tomar, R. S., and A. Jajoo. 2013. "A quick investigation of the detrimental effects of environmental pollutant polycyclic aromatic hydrocarbon fluoranthene on the photosynthetic efficiency of wheat (*Triticum aestivum*)." *Ecotoxicology* 22 (8):1313-8. doi: 10.1007/ s10646-013-1118-1.

Truu, J., M. Truu, Mikk Espenberg, H. Nõlvak, and Jaanis Juhanson. 2015. "Phytoremediation and Plant-Assisted Bioremediation in Soil and Treatment Wetlands: A Review." *The Open Biotechnology Journal* 9:85-92.

Tumaikina, Yu A., O. V. Turkovskaya, and V. V. Ignatov. 2008. "Degradation of hydrocarbons and their derivatives by a microbial association on the base of Canadian pondweed." *Applied Biochemistry and Microbiology* 44 (4):382-388. doi: 10.1134/s00 0368380804008x.

Verane, J., N. C. P. Dos Santos, V. L. da Silva, M. de Almeida, O. M. C. de Oliveira, and I. T. A. Moreira. 2020. "Phytoremediation of polycyclic aromatic hydrocarbons (PAHs) in mangrove sediments using *Rhizophora mangle*." *Mar Pollut Bull* 160:111687. doi: 10.1016/j.marpolbul.2020. 111687.

Wand, H., P. Kuschk, U. Soltmann, and U. Stottmeister. 2002. "Enhanced Removal of Xenobiotics by Helophytes." *Acta Biotechnologica* 22 (1-2):175-181. doi: https://doi.org/10.1002/1521-3846(200205)22:1/2<175::AID-ABIO175>3.0.CO;2-G.

Wang, Z., K. Li, P. Lambert, and C. Yang. 2007. "Identification, characterization and quantitation of pyrogenic polycylic aromatic hydrocarbons and other organic compounds in tire fire products." *J Chromatogr A* 1139 (1):14-26. doi: 10.1016/j.chroma.2006.10.085.

Warren, Jeffrey M., John H. Bassman, John K. Fellman, D. Scott Mattinson, and Sanford Eigenbrode. 2003. "Ultraviolet-B radiation alters phenolic salicylate and flavonoid composition of Populus trichocarpa leaves." *Tree Physiology* 23 (8):527-535. doi: 10.1093/treephys/23.8.527.

Wei, Shiqiang, and Shengwang Pan. 2010. "Phytoremediation for soils contaminated by phenanthrene and pyrene with multiple plant species." *Journal of Soils and Sediments* 10 (5):886-894. doi: 10.1007/s11368-010-0216-4.

Weisman, David, Merianne Alkio, and Adán Colón-Carmona. 2010. "Transcriptional responses to polycyclic aromatic hydrocarbon-induced stress in *Arabidopsis thaliana* reveal the involvement of hormone and defense signaling pathways." *BMC plant biology* 10:59-59. doi: 10.1186/1471-2229-10-59.

Wild, Edward, John Dent, Gareth O. Thomas, and Kevin C. Jones. 2005. "Direct Observation of Organic Contaminant Uptake, Storage, and Metabolism within Plant Roots." *Environmental Science & Technology* 39 (10):3695-3702. doi: 10.1021/es048136a.

Xiao, Nan, Rui Liu, Caixia Jin, and Yuanyuan Dai. 2015. "Efficiency of five ornamental plant species in the phytoremediation of polycyclic aromatic hydrocarbon (PAH)-contaminated soil." *Ecological Engineering* 75:384-391. doi: 10.1016/j.ecoleng.2014.12.008.

Yan, An, Yamin Wang, Swee Ngin Tan, Mohamed Lokman Mohd Yusof, Subhadip Ghosh, and Zhong Chen. 2020. "Phytoremediation: A Promising Approach for Revegetation of Heavy Metal-Polluted Land." *Frontiers in Plant Science* 11 (359). doi: 10.3389/fpls.2020. 00359.

Zezulka, S., M. Kummerova, P. Babula, and L. Vanova. 2013. "*Lemna minor* exposed to fluoranthene: growth, biochemical, physiological and histochemical changes." *Aquat Toxicol* 140-141:37-47. doi: 10.1016/ j.aquatox.2013.05.011.

Zhang, P., and Y. Chen. 2017. "Polycyclic aromatic hydrocarbons contamination in surface soil of China: A review." *Sci Total Environ* 605-606:1011-1020. doi: 10.1016/j.scitotenv.2017.06.247.

Zheng, H., X. Xing, T. Hu, Y. Zhang, J. Zhang, G. Zhu, Y. Li, and S. Qi. 2018. "Biomass burning contributed most to the human cancer risk exposed to the soil-bound PAHs from Chengdu Economic Region, western China." *Ecotoxicol Environ Saf* 159:63-70. doi: 10.1016/j. ecoenv.2018.04.065.

Zhu, Xueqing, Albert D. Venosa, Makram T. Suidan, and Kenneth Lee. 2001. Guidelines for the bioremediation of marine shorelines and freshwater wetlands. US Environmental Protection Agency, Cincinnati.

Zhuang, Jing, Jian Zhang, Xi-Lin Hou, Feng Wang, and Ai-Sheng Xiong. 2014. "Transcriptomic, Proteomic, Metabolomic and Functional Genomic Approaches for the Study of Abiotic Stress in Vegetable Crops." *Critical Reviews in Plant Sciences* 33 (2-3):225-237. doi: 10.1080/07352689.2014.870420.

Chapter 4

Fisheries Contamination Following the NE Brazil Oil Spill: A Case Study of PAHs Levels in Samples from the Fishery Industry

Renato S. Carreira*, Carlos G. Massone,
Wellington Guedes, Ivy de Souza,
Leanderson Coimbra, Otoniel Santana,
Renato Fortes, Lilian Almeida and Arthur L. Scofield
Pontifícia Universidade Católica do Rio de Janeiro, Departamento de Química,
Laboratório de Estudos Marinhos e Ambientais, Rio de Janeiro, RJ, Brazil

1. Introduction

From August 2019 to January 2020, more than 3,000 km of the north-eastern – and a part of the south-eastern – Brazilian coastline were periodically hit by crude oil residues, an accident considered the largest of its kind to ever occur in South Atlantic (Shen et al., 2013; Soares et al., 2020). The oil originated from a spill at sea, but the origin of the spill and those responsible are, to date, still unknown (Zacharias et al., 2021; Lourenço et al., 2020; de Oliveira et al., 2020; Araújo et al., 2021). Even considering the limitations in conducting field sampling and laboratory research imposed by the COVID-19 pandemic (Magalhães et al., 2021), environmental impacts in the aftermath of the oil spill have been demonstrated, including contamination of coastal (mangroves, estuaries, bays, beaches) and marine (coral reefs, seagrass meadows, rodolith beds) areas and associated fauna and flora, and the need to understand long-

---

* Corresponding Author's Email: carreira@puc-rio.br.

In: Polycyclic Aromatic Hydrocarbons
Editor: Warren L. Gregoire
ISBN: 978-1-68507-626-9
© 2022 Nova Science Publishers, Inc.

term impacts in the affected areas is paramount (Magalhães et al., 2021; Araújo et al., 2021; Soares et al., 2021; Nasri Sissini et al., 2020; Magris and Giarrizzo 2020; Lourenço et al., 2020; Escobar 2019; Zacharias et al., 2021; Monteiro et al., 2020; Müller et al., 2021; Campelo et al., 2021; Craveiro et al., 2021). Furthermore, socioeconomic impacts on traditional artisanal fisheries communities and tourist activities have already been mapped, revealing a scenario of increased economic, health and cultural vulnerabilities of the affected communities (Câmara et al., 2021; de Oliveira Estevo et al., 2021).

Specifically considering the threats that oil spills may pose to fisheries contamination, concerns regarding seafood safety and human health are noted (Storelli et al., 2013; Wenzl and Zelinkova 2019), as recently demonstrated following the 2010 DeepWater Horizon (DWH) oil spill in the Gulf of Mexico (Ylitalo et al., 2012; Wickliffe et al., 2018; Simon-Friedt et al., 2016). Polycyclic aromatic hydrocarbons (PAHs) are a class of contaminants of priority attention with respect to seafood safety, as the body-burden of many compounds with carcinogenic effects results from mid- to long-term exposures in water and/or sediments and biomagnification processes (Yender, Michel, and Lord 2002; Wenzl and Zelinkova 2019; Wenzl et al., 2006). As PAH are ubiquitous components in aquatic systems, originating from both petrogenic and pyrolytic sources, any seafood contamination evaluation following oil spills should confirm the specific source of detected PAHs (Boehm, Neff, and Page 2007; Neff 2002; Patin 1999).

To control health risks due to the consumption of contaminated seafood, the 'limit of concern' (LOC) is usually considered. The LOC represents a threshold of maximum PAH concentrations allowed in seafood in risk assessment protocols (Wenzl et al., 2006). Compounds presenting 2 or 3 rings are less toxic in comparison to 4 to 6-ringed PAHs, although a wide range of carcinogenic activities is noted in the latter group (Neff 2002). To adequately compare the toxic effects of PAHs, toxic equivalencies (TEQ) of individual compounds to benzo(a)pyrene are usually considered to establish the LOC in risk analysis methodologies (e.g., Ylitalo et al., 2012). This approach was adopted by the Brazilian Health Regulatory Agency (ANVISA) after the oil spill in NE Brazil, through a Technical Note (ANVISA, 2019). This note considers the consumption of 180 g of fish per day and 60 g of crustaceans and mollusks per day, a five-year period of exposure, and the benzo(a)pyrene (BaPy) equivalent, comprising the weighted sum for eight PAHs with known carcinogenic activity (benzo(a)anthracene, chrysene, benzo(a)fluoranthene, benzo(k)fluoranthene, benzo(a)pyrene, dibenzo(a,h)anthracene, indene(1,2,3-

cd)pyrene and benzo(g,h,i)perylene). The calculated LOC in BaPy-toxic equivalents was 6 µg/kg (or ng/g) for fish and 18 µg/kg (or ng/g) for crustacea and mollusk in the case of the NE Brazil oil spill, which were obtained using Equation 1:

$$LOC = (RL \times BW \times AT \times CF)/(CSF \times CR \times ED) \quad (1)$$

where: LOC is the level of concern; RL is the acceptable risk level, BW is the average body weight of the exposed human population, AT is the average lifespan of the exposed human population, CF is the unit conversion factor, CSF is the cancer slope factor (estimated for individual PAHs according to their carcinogenic potency relative to benzo(a)pyrene), CR is the seafood consumption rate, and ED is the assumed exposure duration (5 years). Reference values for these variables are defined in each risk assessment analysis.

Herein, we present PAH data (2- to 6-ringed, parental and alkylated compounds) obtained from fish and crustacea (lobsters and shrimp) obtained from commercial fishery producers registered at the Brazilian Department of Agriculture, Livestock and Food Supply/Agricultural Defense Secretary (MAPA/SDA). This was an emergency action focused on large producers with the objective to advise Brazilian authorities on the effect of the oil spill upon the regional fish industry and represents the first fisheries samples analyzed in the aftermath of the oil spill.

2. Materials and Methods

2.1. Sampling

A total of 35 samples (11 fishes, 17 shrimps and 7 lobsters) were collected directly from producers registered in the Federal Inspection Service (SIF) from MAPA located in six states (Alagoas, Bahia, Ceará, Paraíba, Pernambuco and Rio Grande do Norte) in the Brazilian NE region. The names of the producers will not be revealed. The samples were collected in October/2019, *i.e.,* when the coastline oiling events took place, and included sites affected by the spill and other sites not necessarily affected. Samples were obtained following a protocol that guarantees their integrity after collection, including wrapping in clean Al-foil, packing, and freezing (–20 °C) until dispatch to the laboratory for analysis. Samples were received at the PUC-Rio Laboratory of Marine and Environmental Studies (LabMAM) and stored frozen (–20 °C). Prior to the analysis, samples were defrosted, and edible fish, shrimp and lobster tissues were sub-sampled in a clean fume-hood and homogenized for further PAH analyses.

2.2. PAH Analyses

The extraction was performed using a pressurized solvent method in tandem with in-cell clean-up purification, as published previously (Massone et al., 2021). The concentrated extracts were further purified by liquid chromatography fractionation on a silica/alumina column. The final extract was fortified with a deuterated PAH mixture (naphthalene-d_8, acenaphene-d_{10}, phenanthrene-d_{10}, chrysene-d_{12} and perylene-d_{12}; 100 ng each) for use as the internal standard for quantification.

PAH identification and quantification followed the EPA-8270D protocol. Briefly, 1 µL of each extract was injected into the equipment, fitted with a DB-5MS column (30 m × 0.25 mm × 0.25 µm) under a constant flow (He at 1.2 mL min^{-1}), and submitted to the following temperature program: 50 °C for 5 min, 50 °C min^{-1} to 80 °C, 6 °C min^{-1} to 280 °C for 20 min, and 12 °C min^{-1} to 305 °C for 10 min. Data acquisition was performed in the selected ion monitoring (SIM) mode, considering typical PAH analysis ions (m/z) (Mauad et al., 2015). Quantification was based on calibration curves containing a mixture of 16 priority PAHs and dibenzothiphene, perylene and benzo(e)pyrene. Alkylated (C1 up to C4) naphthalenes, fluorenes, dibenzothiophenes, anthracenes + phenanthrenes, pyrenes and chrysenes) were quantified considering the response factor of homologs displaying similar structures.

A total of 37 parental and alkylated compounds were identified and quantified (see Table 1 and Figure 1 for details). Typical limits of quantification for individual PAHs were 0.5 ng g^{-1} wet weight (w.w.) for the fish and crustacean samples. A surrogate standard (p-terphenyl-d_{14}) added to each sample (recovery of 76.2 ± 11.1%), as well as the analysis of a certified material (NIST – SRM2976) and a laboratory blank in every sample batch were employed as analytical controls.

2.3. Data Analysis

The PAH fisheries dataset was checked for normality by the Shapiro-Wilk test. As an asymmetric data distribution was observed, significant differences among the variables were evaluated by the non-parametric Kruskal-Wallis and Wilcoxon-Mann-Whitney *post hoc* tests.

Body burden hydrocarbon assignments to petrogenic and/or pyrogenic sources were evaluated based on PAH profiles and diagnostic ratios (Wang, Fingas, and Page 1999). Only 'general' indicators, such as the low- to high-molecular weight ratio (LMW/HML), the percentage of alkylated to non-alkylated compounds and the 'pyrogenic index' according to Wang et al., (1999), were considered. Indicators based on the ratio of specific compounds, such as A/A+Phen or Fl/Fl+Py (Yunker et al., 2002; Santos et al., 2020), were avoided, as the contaminant bioaccumulation process is dependent on animal metabolism and on the chemical properties of each compound, resulting in soft tissue PAH profiles that may not represent the original PAH distribution in the environment (Alegbeleye, Opeolu, and Jackson 2017; Zhao et al., 2014).

3. Results and Discussion

3.1. PAH Occurrence and Composition

A total of 35 samples (11 fishes, 17 shrimp and 7 lobsters) were analyzed concerning their edible PAH tissue concentrations. The median (min/max) concentrations of $\sum$37PAHs (parental and alkylated compounds) were 30.7 ng g^{-1} (12.7 – 52.3 ng g^{-1}) for fish, 35.6 ng g^{-1} (19.2 – 43.0 ng g^{-1}) for lobster and 29.1 ng g^{-1} (9.27 – 197 ng g^{-1}) for shrimp (Table 1; all values on a wet-weight basis). A relatively large difference between the median and mean of $\sum$37PAHs concentrations was noted only for lobster (29.1 and 64.7 ng g^{-1}, respectively), indicating the presence of outlier values in this group of samples. The $\sum$16PAHs represented a higher proportion of total PAHs ($\sum$37 compounds) in fish samples (median of 40.1%), followed by lobster (median of 34.5%) and shrimp (median of 26.5%). The relatively similarity in PAHs levels among the analyzed fisheries was statistically significant ($p < 0.05$) according to the Kruskal-Wallis test.

Limited baseline data concerning PAH contamination of commercial fisheries prior to the oil spill are available. In the aftermath of the spill, one limited dataset on fishes (n = 5) reported $\sum$16PAHs values between 14.3 and 32.1 ng g^{-1} (w.w.) (Soares et al., 2021). A larger set of samples, comprising 194 organisms (143 fishes, 21 crustaceans and 30 mollusks), reported $\sum$37PAHs concentrations ranging from 8.71 to 418 ng g^{-1}, with an overall median of 34.7 ng g^{-1} (Magalhães et al., 2022). All these values are similar to those found in the present study. Regarding other Brazilian regions, data are available, for example, for Guanabara Bay, a tropical waterbody in the largest metropolitan coastal region in the country, with PAHs values ranging from 5-10 ng g^{-1} to up to 320-410 ng g^{-1} for several fishes (Soares-Gomes et al., 2010; da Silva, de Almeida Azevedo, and de Aquino Neto 2007). In the Campos basin, the largest oil producing province in Brazil, fishes caught close to producing-water releasing rigs presented PAHs concentrations ranging from 6.69 to 601 ng g^{-1} (*Caranx crysos*) and from 14.1 to 55.5 ng g^{-1} (*Tylsurus acus*) (Lourenco et al., 2018). Moreover, in estuarine fish from Southern Brazil, Froehner et al., (2018) reported PAHs concentrations ($\sum$16-priority compounds, dry weight) ranging from 26 to 2055 ng g^{-1}.

It is clear that fishes from coastal and marine regions under the influence of chronic exposure to hydrocarbons exhibit higher PAH burdens compared to the samples analyzed herein. However, any comparison of PAH data among different studies must be performed with caution, as the concentrations of each species depend on the balance between exposure (time and type), metabolism capacity, fatty tissue content and intrinsic biological data (age, size, sex, habitat type, and dominant diet, for example). The results are also influenced by the analysis methodology (number of analyzed compounds, normalization unit – dry, wet or lipid weight) employed in each case.

The compositional profile of PAHs is clearly marked by the predominance of low-molecular (LMW) over high-molecular (HMW) weight compounds, with the former accounting for 89.1%, 93.5% and 98.3% of total PAHs for fish, lobster, and shrimp, respectively (median values, see Table 1 for details). The individual PAH profiles reveal a clear predominance of naphthalene (N) and its alkylated analogue (C1N to C3N, but no C4N) homologues, as well the presence of fluorine (F), phenanthrene (Ph), anthracene (A), fluoranthene (Ft) and pyrene (Py), for the three fisheries categories (Figure 1). In fact, alkylated PAHs ranged from 58 to over 73% of total PAHs (Table 1). These results indicate the predominance of petrogenic-derived PAHs in the analyzed samples. Moreover, these data are consistent with the fact that the prevalence of LMW PAHs in animals exposed to oil spills occurs because these

Table 1. Concentrations (ng g^{-1}; wet weight) of total PAHs ($\sum$37PAHs) and the 16-priority US-EPA PAHs ($\sum$16PAHs) in fisheries samples, and PAHs compositional features: ratio of 16 to 37 PAHs, percentage of low-molecular (LMW; 2 and 3-rings parental and alkylated homologues) and high-molecular (HMW; 4 to 6-rings parental and alkylated homologues) weight PAHs, percentage of alkylated PAHs (PAH alk), and the BaP equivalent concentration (see text for details)

		$\sum$37PAHs (ng g^{-1})	$\sum$16PAHs (ng g^{-1})	16PAH / 37PAH (%)	LMW (%)	HMW (%)	PAH alk (%)	BaPy-Equivalent (ng g^{-1})
Fish[*]	Mean	32.5	14.0	40.6	85.8	14.2	59.0	0.32
	St.Dev.	12.8	8.5	10.4	10.9	10.89	11.1	0.75
	Median	30.7	12.9	40.1	89.1	10.9	59.4	0.01
	Min	12.7	3.88	23.5	64.9	<LQ	38.7	<LQ
	Max	52.3	30.1	57.6	100	35.1	76.6	2.44
Lobster[#]	Mean	32.2	11.5	35.5	92.0	8.02	64.4	0.01
	St.Dev.	10.3	3.99	3.88	4.63	4.63	4.04	0.01
	Median	35.6	11.0	34.5	93.6	6.42	65.5	<LQ
	Min	19.2	6.62	31.0	85.9	1.94	58.3	<LQ
	Max	43.0	16.7	41.3	98.1	14.1	69.0	0.04
Shrimp[&]	Mean	64.7	24.5	35.7	90.5	9.51	63.6	0.18
	St.Dev.	63.8	41.5	24.9	21.9	21.9	25.4	0.69
	Median	29.1	9.3	26.5	98.3	1.72	73.3	<LQ
	Min	9.27	3.42	16.1	12.7	<LQ	<LQ	<LQ
	Max	197	178	100	100	87.26	83.9	2.84

[*] Lutjanus synagris, Sparisoma rubripine, Coryphaena hippurus, Epinephelus marginatus, Caranx lugubris, Pseudupeneus maculatus, Lutjanus purpureus.
[#] Panulirus argus, Panulirus laevicauda.
[&] Farfantepenaeus, Xiphopenaeus kroyeri, L. vannamei.

compounds are not as effectively metabolized as HMW PAHs (Romero et al., 2018 and references therein). A slightly different output, however, was obtained based on the index proposed by Wang et al., (1999), which considers the sum of 3-6 ringed non-alkylated PAHs to the sum of 5-series alkylated PAHs (naphthalenes, phenanthrene, dibenzothiophene, fluorene and chrysene). According to the plot presented (Figure 2), fish and lobster samples seem to be exposed to a mixture of petrogenic and pyrogenic PAHs, while the shrimp samples accumulated petrogenic PAHs. Therefore, we cannot rule out the influence of chronic exposure – although at low levels, based on the $\sum 37$PAHs concentrations, as discussed previously – to the PAH body burden in the fisheries considered in the present study.

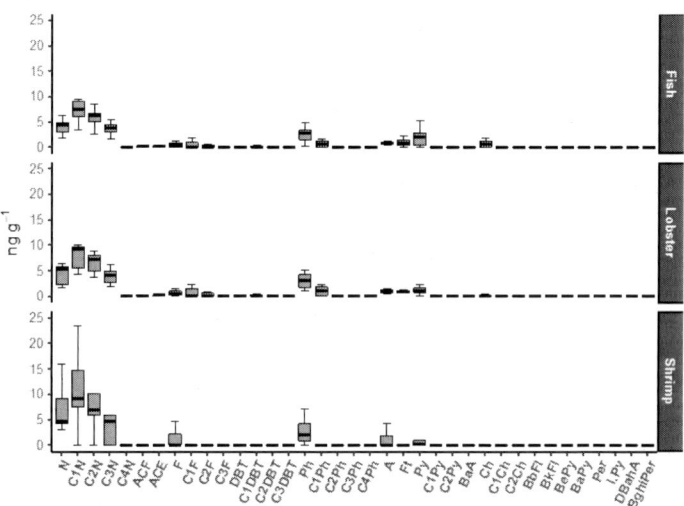

Figure 1. PAH concentration (means ± standard deviations) profiles in fishes, shrimp and lobsters from the fish industry from NE Brazil folowing the 2019 oil spill. Legend: naphthalene (N), C1-naphthalenes (C1N), C2-naphthalenes (C2N), C3-naphthalenes (C3N), C4-naphthalenes (C4N), acenaphthylene (ACF), acenaphthene (ACE), fluorine (F), C1-fluorines (C1F), C2-fluorines (C2F), C3-fluorines (C3F), dibenzothiophene (DBT), C1-dibenzothiophenes (C1DBT), C2-dibenzothiophenes (C2DBT), C3-dibenzothiophenes (C3DBT), phenanthrene (Ph), C1-phenanthrenes (C1Ph), C2-phenanthrenes (C2Ph), C3-phenanthrenes (C3Ph), C4-phenanthrenes (C4Ph), anthracene (A), fluoranthene (Fl), pyrene (Py), C1-pyrenes (C1Py), C2-pyrenes (C2Py), benzo[a]anthracene (BaA), chrysene (Ch), C1-chrysenes (C1Ch), C2-chrysenes (C2Ch), benzo[b]fluoranthene (BbFl), benzo[k]fluoranthene (BkFl), benzo[e]pyrene (BePy), benzo[a]pyrene (BaPy), perylene (Per), indeno[1,2,3-cd]pyrene (I-Py), dibenz[a,h]anthracene (DBahA) and benzo[ghi]perylene (BghiPer).

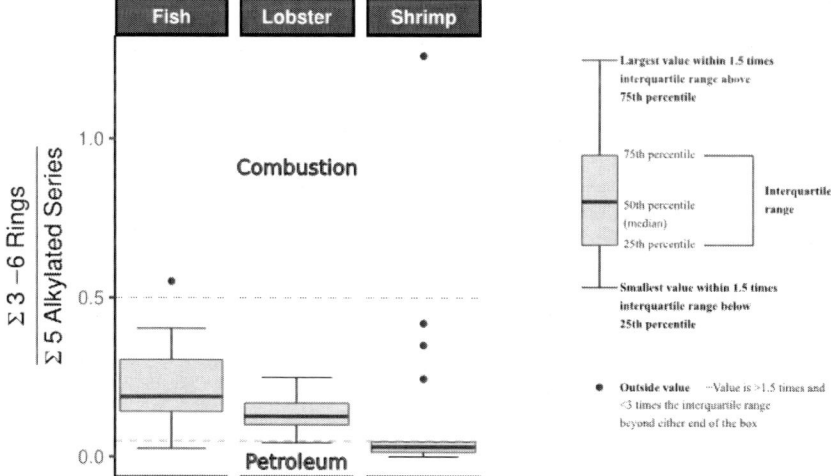

Figure 2. Box-plots (median, percentiles and outliers) of the relative ratio of $\sum$(other 3–6-ring PAHs)/$\sum$(5-alkylated PAHs) obtained for the fish, lobster and shrimp samples. Petrogenic PAHs fall in the range below 0.05, whereas pyrogenic PAHs are above 0.5. Values between the two thresholds represent a mixture of sources.

3.2. PAHs Risk Analysis Concerning Seafood Safety

The recognized toxic potential of PAHs justifies concerns regarding seafood safety following oil spills (Gohlke et al., 2011). The most employed approach is the consideration of a limit of concern (LOC) – a threshold calculated based on the relative toxicities of carcinogenic PAHs relative to benzo(a)pyrene – in risk analysis protocols, which must also consider intrinsic factors of a given population, such as age and consumption patterns (Yender, Michel, and Lord 2002; Wenzl and Zelinkova 2019).

The Brazilian National Health Surveillance Agency (ANVISA), through a technical note (NT27/2019/SEI/GGALI/DIRE2/ANVISA), proposed the use of BaPy equivalent as the LOC to evaluate the quality of seafood harvested along over 3,000 km of coastline affected by the 2019-2020 oil spill in NE Brazil and the following thresholds: 6 ng/g of BaPy-equivalent for fish and 18 ng/g BaPy-equivalent for crustacea and mollusks.

All the fisheries samples analyzed in the present study ranked below the LOC stablished by ANVISA to guarantee seafood quality (Figure 3 and Table 1)

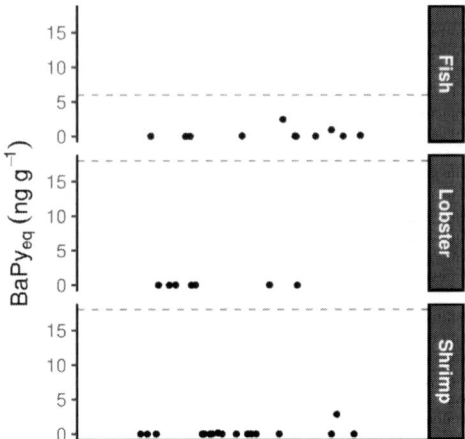

Figure 3. BaPy equivalent values for the commercial fisheries samples obtained in in October/2019 from registered producers. The limit of concern (LOC) in BaPy-equivalent values is indicated by a red line for each taxonomic group (6 ng/g of BaPy-equivalent for fish and 18 ng/g BaPy-equivalent for lobsters and shrimp).

Conclusion

The commercial fisheries samples (fish, lobster, and shrimp) collected directly from registered producers located in several states in NE Brazil whose coastlines were contaminated by oil residues from a spill exhibited relatively low $\sum37$PAHs concentrations (median of 30.7 ng g^{-1}; average of 67.4 ± 59.2 ng g^{-1}). These values are similar to values detected in other fisheries from different areas affected by the spill (Soares et al., 2021; Magalhães et al., 2022), but are at least one order of magnitude lower than determined in meso-pelagic fish sampled after the 2010 Deep-Water Horizon blowout in the Gulf of Mexico (Romero et al., 2018). Moreover, none of the 35 samples analyzed herein exhibit PAHs concentrations – in terms of BaPy-equivalent values – above the limit of concern proposed for the risk analysis of seafood safety specific to the NE Brazil oil spill case. The data presented herein are relative to the first fisheries samples analyzed in the aftermath of the NE Brazil oil spill. Although these results cannot be used as a rule, they indicate that the commercial fisheries samples considered herein are safe for human consumption.

Acknowledgments

The authors thank the public servers from the Brazilian Department of Agriculture, Livestock and Food Supply/Agricultural Defense Secretary (MAPA/SDA) for providing the samples and for approving the publication of the data presented herein.

References

Alegbeleye, Oluwadara Oluwaseun, Beatrice Oluwatoyin Opeolu, and Vanessa Angela Jackson. 2017. "Polycyclic Aromatic Hydrocarbons: A Critical Review of Environmental Occurrence and Bioremediation." *Environmental Management* 60 (4): 758-783.

Araújo, Kelvin C., Matheus C. Barreto, Alcides S. Siqueira, Anne Caroline P. Freitas, Levi G. Oliveira, Maria Eugênia P. A. Bastos, Maria Eduarda P. Rocha, Lucimary A. Silva, and Wallace D. Fragoso. 2021. "Oil spill in northeastern Brazil: Application of fluorescence spectroscopy and PARAFAC in the analysis of oil-related compounds." *Chemosphere* 267: 129154.

Boehm, P. D., J. M. Neff, and D. S. Page. 2007. "Assessment of polycyclic aromatic hydrocarbon exposure in the waters of Prince William Sound after the Exxon Valdez oil spill: 1989-2005." *Marine Pollution Bulletin* 54 (3): 339-356.

Câmara, Samuel Façanha, Francisco Roberto Pinto, Felipe Roberto da Silva, Marcelo de Oliveira Soares, and Thiago Matheus De Paula. 2021. "Socioeconomic vulnerability of communities on the Brazilian coast to the largest oil spill (2019–2020) in tropical oceans." *Ocean & Coastal Management* 202: 105506.

Campelo, Renata Polyana de Santana, Cynthia Dayanne Mello de Lima, Claudeilton Severino de Santana, Alef Jonathan da Silva, Sigrid Neumann-Leitão, Beatrice Padovanni Ferreira, Marcelo de Oliveira Soares, Mauro de Melo Júnior, and Pedro Augusto Mendes de Castro Melo. 2021. "Oil spills: The invisible impact on the base of tropical marine food webs." *Marine Pollution Bulletin* 167: 112281.

Craveiro, Nykon, Rodrigo Vinícius de Almeida Alves, Juliana Menezes da Silva, Edson Vasconcelos, Flavio de Almeida Alves-Junior, and José Souto Rosa Filho. 2021. "Immediate effects of the 2019 oil spill on the macrobenthic fauna associated with macroalgae on the tropical coast of Brazil." *Marine Pollution Bulletin* 165: 112107.

da Silva, Taís Freitas, Débora de Almeida Azevedo, and Francisco Radler de Aquino Neto. 2007. "Polycyclic Aromatic Hydrocarbons in Fishes and Sediments from the Guanabara Bay, Brazil." *Environmental Forensics* 8 (3): 257-264.

de Oliveira Estevo, Mariana, Priscila F. M. Lopes, José Gilmar Cavalcante de Oliveira Júnior, André Braga Junqueira, Ana Paula de Oliveira Santos, Johnny Antonio da Silva Lima, Ana Claudia Mendes Malhado, Richard J. Ladle, and João Vitor Campos-Silva. 2021. "Immediate social and economic impacts of a major oil spill on Brazilian coastal fishing communities." *Marine Pollution Bulletin* 164: 111984.

de Oliveira, Olívia M. C., Antônio F. de S. Queiroz, José R. Cerqueira, Sarah A. R. Soares, Karina S. Garcia, Aristides Pavani Filho, Maria de L. da S. Rosa, Caroline M. Suzart, Liliane de L. Pinheiro, and Ícaro T. A. Moreira. 2020. "Environmental disaster in the northeast coast of Brazil: Forensic geochemistry in the identification of the source of the oily material." *Marine Pollution Bulletin* 160: 111597.

Escobar, Herton. 2019. "Mystery oil spill threatens marine sanctuary in Brazil." *Science* 366 (6466): 672.

Froehner, Sandro, Juliane Rizzi, Luciane Maria Vieira, and Juan Sanez. 2018. "PAHs in Water, Sediment and Biota in an Area with Port Activities." *Archives of Environmental Contamination and Toxicology* 75 (2): 236-246.

Gohlke, Julia M., Dzigbodi Doke, Meghan Tipre, Mark Leader, and Timothy Fitzgerald. 2011. "A Review of Seafood Safety after the Deepwater Horizon Blowout." *Environmental Health Perspectives* 119 (8): 1062-1069.

Lourenco, R. A., E. Francioni, Ahmft da Silva, C. A. Magalhaes, F. D. C. Gallotta, F. F. de Oliveira, J. M. de Souza, L. F. M. de Araujo, L. P. de Araujo, M. A. G. de Araujo Junior, M. de Fatima Guadalupe Meniconi, and M. A. F. de Souza Bindes Gomes Lopes. 2018. "Bioaccumulation Study of Produced Water Discharges from Southeastern Brazilian Offshore Petroleum Industry Using Feral Fishes." *Arch. Environ. Contam. Toxicol.* 74 (3): 461-470.

Lourenço, Rafael André, Tatiane Combi, Marcelo Da Rosa Alexandre, Silvio Tarou Sasaki, Eliete Zanardi-Lamardo, and Gilvan Takeshi Yogui. 2020. "Mysterious oil spill along Brazil's northeast and southeast seaboard (2019–2020): Trying to find answers and filling data gaps." *Marine Pollution Bulletin* 156: 111219.

Magalhães, Karine Matos, Kcrishna Vilanova de Souza Barros, Maria Cecília Santana de Lima, Cristina de Almeida Rocha-Barreira, José Souto Rosa Filho, and Marcelo de Oliveira Soares. 2021. "Oil spill + COVID-19: A disastrous year for Brazilian seagrass conservation." *Science of The Total Environment* 764: 142872.

Magalhães, Karine Matos, Carreira, Renato, Rosa Filho, José Souto, Rocha, Pedro Palmeira, Santana, Francisco Marcante and Yogui, Gilvan Takeshi. "Polycyclic aromatic hydrocarbons (PAHs) in fishery resources affected by the 2019 oil spill in Brazil: short-term environmental health and seafood safety." *Marine Pollution Bulletin*, 175: 113334.

Magris, Rafael Almeida, and Tommaso Giarrizzo. 2020. "Mysterious oil spill in the Atlantic Ocean threatens marine biodiversity and local people in Brazil." *Marine Pollution Bulletin* 153: 110961.

Massone, C. G., A. A. Santos, P. G. Ferreira, and R. S. Carreira. 2021. "A baseline evaluation of PAH body burden in sardines from the southern Brazilian shelf." *Marine Pollution Bulletin* 163: 111949.

Mauad, Cristiane R., Angela de L. R. Wagener, Carlos G. Massone, Mayara da S. Aniceto, Letícia Lazzari, Renato S. Carreira, and Cássia de O. Farias. 2015. "Urban rivers as conveyors of hydrocarbons to sediments of estuarine areas: Source characterization, flow rates and mass accumulation." *Science of The Total Environment* 506–507 (0): 656-666.

Monteiro, Caroline Barbosa, Phelype Haron Oleinik, Thalita Fagundes Leal, Wiliam Correa Marques, João Luiz Nicolodi, and Bruna de Carvalho Faria Lima Lopes. 2020.

"Integrated environmental vulnerability to oil spills in sensitive areas." *Environmental Pollution* 267: 115238.

Müller, Marius Nils, Gilvan Takeshi Yogui, Alfredo Olivera Gálvez, Luiz Gustavo de Sales Jannuzzi, Jesser Fidelis de Souza Filho, Manuel de Jesus Flores Montes, Pedro Augusto Mendes de Castro Melo, Sigrid Neumann-Leitão, and Eliete Zanardi-Lamardo. 2021. "Cellular accumulation of crude oil compounds reduces the competitive fitness of the coral symbiont Symbiodinium glynnii." *Environmental Pollution* 289: 117938.

Nasri Sissini, M., F. Berchez, J. Hall-Spencer, N. Ghilardi-Lopes, V. F. Carvalho, N. Schubert, G. Koerich, G. Diaz-Pulido, J. Silva, E. Serrão, J. Assis, R. Santos, S. R. Floeter, L. Rörig, J. B. Barufi, A. F. Bernardino, R. Francini-Filho, A. Turra, L. C. Hofmann, J. Aguirre, L. Le Gall, V. Peña, M. C. Nash, S. Rossi, M. Soares, G. Pereira-Filho, F. Tâmega, and P. A. Horta. 2020. "Brazil oil spill response: Protect rhodolith beds." *Science* 367 (6474): 156.

Neff, J. M. 2002. *Bioaccumulation in marine organisms: effects of contaminants from oil well produced water.* Amsterdam: Elsevier.

Patin, S. 1999. *Environmental impact of the offshore oil and gas industry.* Translated by Elena Cascio. New York: EcoMonitor Publishing.

Romero, Isabel C., Tracey Sutton, Brigid Carr, Ester Quintana-Rizzo, Steve W. Ross, David J. Hollander, and Joseph J. Torres. 2018. "Decadal Assessment of Polycyclic Aromatic Hydrocarbons in Mesopelagic Fishes from the Gulf of Mexico Reveals Exposure to Oil-Derived Sources." *Environmental Science & Technology* 52 (19): 10985-10996.

Santos, Felipe R., Patricia A. Neves, Bianca S. M. Kim, Satie Taniguchi, Rafael A. Lourenço, Cristian T. Timoszczuk, Basílio M. T. Sotão, Rosalinda C. Montone, Rubens C. L. Figueira, Michel M. Mahiques, and Márcia C. Bícego. 2020. "Organic contaminants and trace metals in the western South Atlantic upper continental margin: Anthropogenic influence on mud depocenters." *Marine Pollution Bulletin* 154: 111087.

Shen, Huizhong, Ye Huang, Rong Wang, Dan Zhu, Wei Li, Guofeng Shen, Bin Wang, Yanyan Zhang, Yuanchen Chen, Yan Lu, Han Chen, Tongchao Li, Kang Sun, Bengang Li, Wenxin Liu, Junfeng Liu, and Shu Tao. 2013. "Global Atmospheric Emissions of Polycyclic Aromatic Hydrocarbons from 1960 to 2008 and Future Predictions." *Environmental Science & Technology* 47 (12): 6415-6424.

Simon-Friedt, Bridget R., Jessi L. Howard, Mark J. Wilson, David Gauthe, Donald Bogen, Daniel Nguyen, Ericka Frahm, and Jeffrey K. Wickliffe. 2016. "Louisiana residents' self-reported lack of information following the Deepwater Horizon oil spill: Effects on seafood consumption and risk perception." *Journal of Environmental Management* 180: 526-537.

Soares, Emerson Carlos, Mozart Daltro Bispo, Vivian Costa Vasconcelos, João Inácio Soletti, Sandra Helena Vieira Carvalho, Maria Janaína de Oliveira, Mayara Costa dos Santos, Emerson dos Santos Freire, Aryanna Sany Pinto Nogueira, Francisco Antônio da Silva Cunha, Rafael Donizete Dutra Sandes, Raquel Anne Ribeiro dos Santos, Maria Terezinha Santos Leite Neta, Narendra Narain, Carlos Alexandre Borges Garcia, Silvânio Silvério Lopes da Costa, and Josué Carinhanha Caldas Santos. 2021.

"Oil impact on the environment and aquatic organisms on the coasts of the states of Alagoas and Sergipe, Brazil - A preliminary evaluation." *Marine Pollution Bulletin* 171: 112723.

Soares, M. O., C. E. P. Teixeira, L. E. A. Bezerra, S. Rossi, T. Tavares, and R. M. Cavalcante. 2020. "Brazil oil spill response: Time for coordination." *Science* 367 (6474): 155.

Soares-Gomes, Abílio, Roberta L. Neves, Ricardo Aucélio, Paulo H. Van Der Ven, Fábio B. Pitombo, Carla L. T. Mendes, and Roberta L. Ziolli. 2010. "Changes and variations of polycyclic aromatic hydrocarbon concentrations in fish, barnacles and crabs following an oil spill in a mangrove of Guanabara Bay, Southeast Brazil." *Marine Pollution Bulletin* 60 (8): 1359-1363.

Storelli, Maria Maddalena, Grazia Barone, Veronica Giuliana Perrone, and Arianna Storelli. 2013. "Risk characterization for polycyclic aromatic hydrocarbons and toxic metals associated with fish consumption." *Journal of Food Composition and Analysis* 31 (1): 115-119.

Wang, Zhendi, Merv Fingas, and David S. Page. 1999. "Oil spill identification." *Journal of Chromatography A* 843 (1-2): 369-411.

Wenzl, Thomas, Rupert Simon, Elke Anklam, and Juliane Kleiner. 2006. "Analytical methods for polycyclic aromatic hydrocarbons (PAHs) in food and the environment needed for new food legislation in the European Union." *TrAC Trends in Analytical Chemistry* 25 (7): 716-725.

Wenzl, Thomas, and Zuzana Zelinkova. 2019. "Polycyclic Aromatic Hydrocarbons in Food and Feed." In *Encyclopedia of Food Chemistry*, edited by Laurence Melton, Fereidoon Shahidi and Peter Varelis, 455-469. Oxford: Academic Press.

Wickliffe, Jeffrey K., Bridget Simon-Friedt, Jessi L. Howard, Ericka Frahm, Buffy Meyer, Mark J. Wilson, Deepa Pangeni, and Edward B. Overton. 2018. "Consumption of Fish and Shrimp from Southeast Louisiana Poses No Unacceptable Lifetime Cancer Risks Attributable to High-Priority Polycyclic Aromatic Hydrocarbons." *Risk Analysis* 38 (9): 1944-1961.

Yender, R., J. Michel, and C. Lord. 2002. *Managing seafood safety after an oil spill*. Office of Response and Restoration, National Oceanic and Atmospheric Administration (Seatle).

Ylitalo, Gina M., Margaret M. Krahn, Walton W. Dickhoff, John E. Stein, Calvin C. Walker, Cheryl L. Lassitter, E. Spencer Garrett, Lisa L. Desfosse, Karen M. Mitchell, Brandi T. Noble, Steven Wilson, Nancy B. Beck, Ronald A. Benner, Peter N. Koufopoulos, and Robert W. Dickey. 2012. "Federal seafood safety response to the Deepwater Horizon oil spill." *Proceedings of the National Academy of Sciences of the United States of America* 109 (50): 20274-20279.

Yunker, Mark B., Robie W. Macdonald, Roxanne Vingarzan, Reginald H. Mitchell, Darcy Goyette, and Stephanie Sylvestre. 2002. "PAHs in the Fraser River basin: a critical appraisal of PAH ratios as indicators of PAH source and composition." *Organic Geochemistry* 33: 489-515.

Zacharias, Daniel Constantino, Carine Malagolini Gama, Joseph Harari, Rosmeri Porfirio da Rocha, and Adalgiza Fornaro. 2021. "Mysterious oil spill on the Brazilian coast –

Part 2: A probabilistic approach to fill gaps of uncertainties." *Marine Pollution Bulletin* 173: 113085.

Zhao, Zhonghua, Lu Zhang, Yongjiu Cai, and Yuwei Chen. 2014. "Distribution of polycyclic aromatic hydrocarbon (PAH) residues in several tissues of edible fishes from the largest freshwater lake in China, Poyang Lake, and associated human health risk assessment." *Ecotoxicology and Environmental Safety* 104: 323-331.

Chapter 5

Associations between Biliary Polycyclic Aromatic Hydrocarbons, Biomorphometric Indices and Biliverdin as a Feeding Status Proxy in Mullet (*Mugil liza*) from a Chronically Contaminated Estuary in Southeastern Brazil

Rachel Ann Hauser-Davis[1,*]**,**
Roberta Lyrio Santos Neves[2]**,**
Danielle Lopes Mendonça[2]
and Roberta Lourenço Ziolli[2]

[1]Laboratório de Avaliação e Promoção da Saúde Ambiental,
Instituto Oswaldo Cruz (Fiocruz), Rio de Janeiro, RJ, Brazil
[2]Departamento de Química,
Pontifícia Universidade Católica do Rio de Janeiro (PUC-Rio),
Rio de Janeiro, RJ, Brazil
[3]Instituto de Biociências,
Universidade Federal do Estado Rio de Janeiro (UNIRIO),
Rio de Janeiro, RJ, Brazil

---

[*] Corresponding Author's Email: rachel.hauser.davis@gmail.com.

In: Polycyclic Aromatic Hydrocarbons
Editor: Warren L. Gregoire
ISBN: 978-1-68507-626-9
© 2022 Nova Science Publishers, Inc.

Abstract

Polycyclic aromatic hydrocarbons (PAH) are rapidly metabolized and excreted by vertebrates, resulting in low PAH concentrations in tissues routinely used in biomonitoring efforts, such as muscle and liver. Therefore, alternatives have been applied in this regard, such as the assessment of biliary PAH concentrations and their metabolites. However, fish feeding status has been reported as affecting PAH metabolite concentrations and is still poorly understood. Furthermore, it is known that PAH may significantly affect piscine biomorphometric indices, although studies are still scarce in this regard. Therefore, the aim of this study was to evaluate potential associations between biliary polycyclic aromatic hydrocarbons (PAH), biomorphometric indices and biliverdin applied as a feeding status proxy in mullet (*Mugil liza*) from a chronically contaminated estuary in Southeastern Brazil.

Keywords: biomonitoring, oil spill, environmental contamination, organic pollutants, fish

1. Introduction

Polycyclic aromatic hydrocarbons (PAH) are ubiquitous environmental contaminants comprising organic compounds containing two or more fused aromatic rings (Kim et al., 2013). The presence of these contaminants in the environment, especially aquatic ecosystems, is of concern, since they display carcinogenic and mutagenic properties and can bioaccumulate in aquatic organisms, affecting many trophic components (Honda et al., 2020).

Major PAH input sources in the marine environment comprise anthropogenic activities, such as fossil fuel burning, oil spills, urban runoff and industrial and urban discharges (Latimer Zheng, 2003). Atmospheric deposition is also an important source of environmental PAH contamination, through incomplete fossil fuel combustion by motor vehicles, considered the main pyrogenic PAH contamination source (Fänrich, Guilbault & Pravda, 2002; Dallarosa, et al., 2005). Natural sources, which are usually minor, include plant biomass combustion, forest fires and the diagenesis of natural precursors (Abas & Mohamad, 2011).

Vertebrates exposed to PAH show very fast uptake and metabolization of these xenobiotics from their diet or the environment. As PAH compounds are mostly insoluble in aqueous media, to enhance excretion through urine or bile, they must be metabolized and transformed into more water-soluble metabolites (McCarthy & Sinal, 2005). These authors indicate that the biotransformation process generally involves two distinct stages, referred to as Phase I and Phase II reactions. In Phase I, biotransformation occurs through reduction, hydrolysis and/or oxidation reactions, which introduce functional groups in the xenobiotic compounds. The main reaction at this stage of the process is oxidation, catalyzed mainly by cytochrome P-450 monooxygenase, monoamino oxidases and flavin monooxygenases. Hydrolysis reactions are catalyzed mostly by epoxide hydrolases, peptidases and A-esterases, while reductions are catalyzed by various enzymes, including carbonyl reductases and glutathione reductases, as well as by non-enzymatic processes through reducing agents. Phase II often involves conjugation reactions, in an attempt to neutralize the offending xenobiotic, catalyzed by glutathione S-transferases, uridine diphosphoglycuronosyl transferases and sulfotransferases. In general, the increased polarity of the xenobiotic molecule following the biotransformation process results in less toxic compounds, leading to detoxification, although metabolic activation, where the xenobiotic is converted into a more toxic compound, has been known to occur in some cases (McCarthy & Sinal, 2005).

Bile, in addition to playing an important role in fat digestion and absorption, aiding in the emulsion of large fat molecules, plays a significant role in xenobiotic excretion in several animal groups (Boyer, 2013). Bile xenobiotic elimination assessments were scarce during the first half of the twentieth century, but from the second half onwards, with the appearance of a wide variety of synthetic chemical substances, the importance of bile as a way of excreting these compounds was recognized. Concerning PAH, pyrene, a 4-ringed compound, is regarded the best general indicator of PAH exposure in fish, as it is the major metabolite found in the bile of fish individuals exposed to PAH contamination (Van der Oost, 2003).

Both biliverdin and bile protein concentrations are indicative of fish dietary status, with higher concentrations indicating concentrated bile, the result of starving periods between feeding events (Richardson et al., 2004). As biliary concentration also leads to contaminant concentration, some studies have recommended normalizing pollutant level data in relation to biliverdin concentrations, in an attempt to reduce data set variance concerning differential feeding statuses.

Some simple piscine morphometric indices have been used alongside contaminant data in several fish species to verify potential growth and health effects. These include the hepatosomatic index (HSI), which identifies possible liver disorders and may reflect the reproduction period of a given species, and the condition factor (CF) that assesses general fish health (Van der Oost, et al., 2002).

In this context, this study aimed to verify potential correlations between fish biliary PAH in the form of 1-OH-pyrene equivalents, piscine biomorphometric indices and biliverdin as a feeding status proxy in mullet (*Mugil liza*) sampled from a chronically contaminated PAH estuary in Southeastern Brazil.

2. Methodology

2.1. Study Area

Guanabara Bay (GB) is located in the metropolitan region of the state of Rio de Janeiro and home to over 8 million people, about two thirds of the entire metropolitan Rio de Janeiro region population. The climate at GB is humid tropical with hot and rainy summers and cold, dry winters, and the entire area undergoes significant marine influence, due to a local upwelling phenomenon. The GB is characterized by both a salinity and temperature gradient, decreasing from bay's mouth near the ocean to the inner areas of the bay. The temperature increases from the entrance of the bay to the bottom, in response to the advection of colder sea water. In addition, a decreasing pollution gradient toward the Atlantic Ocean is also verified (Fistarol et al., 2015).

Over 17,000 industries, including pharmaceutical industries and oil refineries, as well as oil and gas terminals, shipyards and commercial ports, surround the bay, resulting in immense amounts of contamination inputs. These inputs have been estimated as follows: around 150 t of industrial wastewater, 18 t day^{-1} of petroleum hydrocarbons and 813 t day^{-1} of solid waste (Fistarol et al., 2015). However, despite these high chronic inputs the bay is still bordered by about 90 km^2 of mangroves and maintains high fishery productivity and ecological importance, as mangroves are essential as feeding, protection, and reproduction sites for many animal species (Fistarol et al., 2015, Prestelo & Vianna, 2016).

2.2. Fish Species

Fish belonging to the Mugilidae family display a wide geographic distribution, occurring in tropical and subtropical waters worldwide, especially in coastal estuarine regions (Nelson, 2006). These fish are detritivorous, feeding on organic material derived from bodies of dead organisms and excretions left by living organisms, and also on a wide variety of invertebrates that can occur in this detritus (Nelson, 2006). Their detritivorous feeding behavior makes them more vulnerable to pollution, as many contaminants are often present in high concentrations in sediments and organic material present throughout the water column (Mansour & Sidky, 2002; Claro et al., 2014). In Brazil, *Mugil liza* is an estuarine-dependent species comprising an important fishing resource, with fisheries landings estimated as between 164 to 1020 tons between 2011 and 2018 in the state of Rio de Janeiro, according to the state's Institute of Fisheries Foundation (da Costa et al., 2021).

2.3. Fish Sampling and Bile Collection

Fish were randomly obtained from artisanal fishers that fish throughout the GB using both gillnets and corrals in Autumn 2007 (Figure 1) and transported to the laboratory in sealed Styrofoam boxes containing ice. Upon arrival, each fish was weighed and measured to the nearest gram and centimeter using a semi-analytical balance (Marte, São Paulo) and an icthyometer. After dissection, livers were weighed whole and bile was obtained with the aid of sterile syringes, transferred to 2 mL microtubes. All samples were at -20°C until analyses.

2.4. Morphobiometric Index Calculations

The hepatosomatic index (HSI) and Fulton's Condition Factor (FCF) were calculated as FCF = 100 W_T/L_T^3, where W_T, in grams comprises total fish weight and L_T, in cm consists in total fish length, and HIS = 100 W_L/W_T, where W_L, in grams, comprises fish liver weight and W_T, in cm consists of fish weight (W_T, in cm) (Maddock and Burton, 1998).

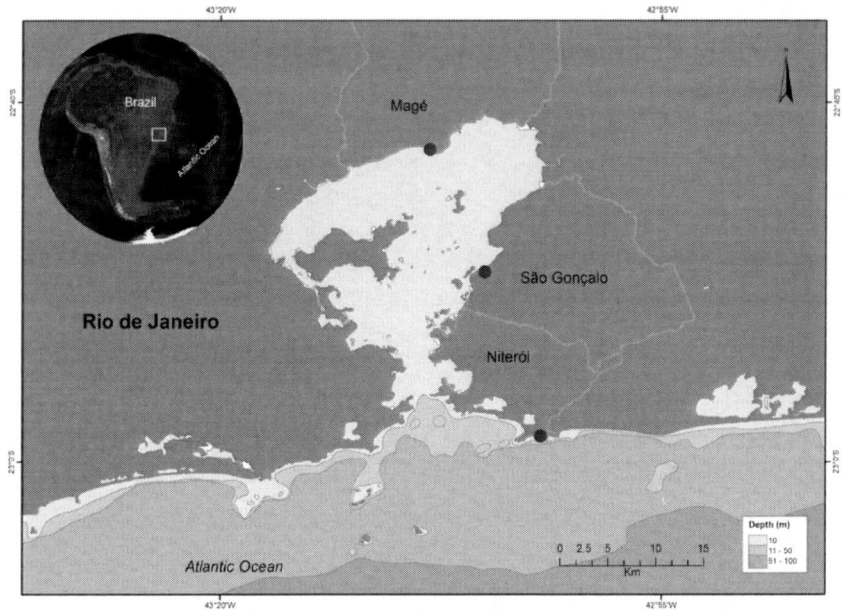

Figure 1. *Mugil liza* sampling sites located throughout Guanabara Bay, in Rio de Janeiro, Southeastern Brazil.

2.5. Biliverdin (BV) Determinations

For biliverdin determinations, bile samples were diluted in 48% ethanol (1:100 v/v) and absorbances were determined at 380 nm, the maximum absorbance peak for biliverdin, using a Perkin Elmer Lambda 19 UV-Vis spectrophotometer, according to previous studies (Aas et al., 2000; Richardson et al., 2004). Biliverdin concentrations were calculated using analytical curves prepared using biliverdin hydrochloride (97.0% purity, Sigma-Aldrich, São Paulo) as an external standard. All analyses were performed in triplicate.

2.6. Total Fluorescent Compound (TFC) Determinations

For Total fluorescent compound (TFC) determinations, bile samples were analyzed according to Beyer et al., (1996) and Aas et al., (1998), by simple dilution in 48% ethanol (1:2000 v/v). Fluorescence readings were performed on a Perkin Elmer LS 55 Luminescence Spectrometer using quartz cuvettes. The optimal1-hydroxy-pyrene excitation/emission wavelength pair used to detect 1-hydroxy-pyrene was previously selected by screening using a 1-hydroxy-pyrene standard (97.0% purity, St. Louis, USA) diluted in 48% ethanol (1:2000 v/v), set as 345/387 nm. As other PAH can fluoresce in the same wavelength pair, the results were expressed as 1-hydroxy-pyrene equivalents. All analyses were performed in triplicate.

2.7. Statistical Analyses

Data normality was tested by the Shapiro-Wilk test. As the data were non-normally distributed, a non-parametric Mann-Whitney test was applied to assess differences between the means of each of the assessed variables (CF, HSI, BV, TFC) for male and female *M. liza* specimens sampled from Guanabara Bay. A Principal Component Analysis was carried out to identify statistical associations between the investigated variables following data normalization by the Z score method. Only functions presenting $p < 0.05$ and eigenvalues above 1 according to Kaiser's rule were considered statistically significant.

3. Results and Discussion

A total of 29 fish were sampled, 13 females and 16 males. The morphobiometric *M. liza* data and calculated indices (FCF and HSI) are displayed in Table 1. Data are presented as the means ± standard deviations.

Table 1. Morphobiometric data for mullet (*M. liza*, n = 29) specimens sampled from Guanabara Bay, southeastern Brazil

aZ	Weight (g)	TL (cm)	CF	HSI
Minimum	313.60	34.40	0.69	1.15
Maximum	2,500.00	64.00	1.12	2.90
Mean ± SD	1,036.00 ± 533.20	48.31 ± 8.74	0.84 ± 0.10	1.79 ± 0.46

The CF ranged from 0.69 to 1.12 and the HIS, from 1.15 to 2.90 when grouping all samples regardless of sex. Statistically significant differences (p

< 0.05), however were observed between males and females for both the CF, which ranged from 0.71 to 1.12 in females and 0.69 to 0.89 in males, and the HIS, which ranged from 1.41 to 2.9 in females and 1.15 to 2.04 in males, respectively (Figure 2).

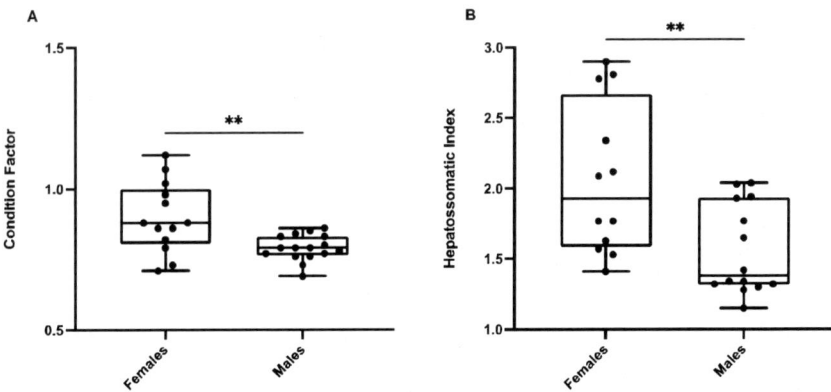

Figure 2. Box plot graphs representing the (A) Condition Factor and (B) Hepatossomatic index values in male and female mullet (*M. liza*, n = 29) specimens sampled from Guanabara Bay, southeastern Brazil. Asterisks indicate significant differences between groups. The horizontal lines within the box plots display the medians, the upper line indicates the 75th quartile and the lower line, the 25th quartile. The whiskers indicate the maximum and minimum ranges. Each individual specimen is indicated by a dot within the boxes.

All fish were sampled during the same year and season, to avoid seasonality differences and consequent reproductive effort interferences among individuals. Therefore, the significant differences ($p < 0.05$) observed between males and females for FCF and HSI, in which females present higher means for both morphometric indices, seem to indicate the proximity of the reproductive season, reported as from March to August at GB (da Costa et al., 2021), and energy mobilization towards vitellogenin production in females, which is synthesized in the female liver as the egg yolk precursor, as fish ranged from slightly below the sexual maturity length reported in the literature of 40 cm to 64 cm for several Brazilian estuarine areas, indicating mostly mature individuals (Esper, Menezes and Esper, 2000; Albieri & Araújo, 2010; Lemos et al., 2014; da Costa et al., 2021).

The BV and TFC results for *M. liza* bile are displayed in Table 2. Data are presented as means ± standard deviation.

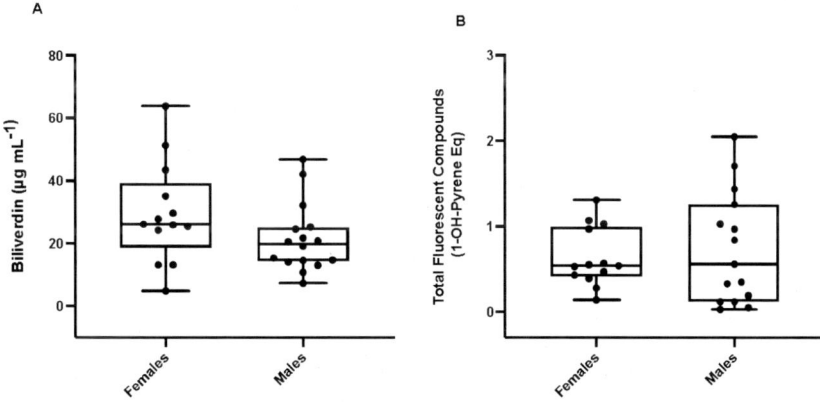

Figure 3. Box plot graphs representing (A) biliverdin concentrations and (B) total fluorescent compounds values in male and female mullet (*M. liza*, n = 29) specimens sampled from Guanabara Bay, southeastern Brazil. Asterisks indicate significant differences between groups. The horizontal lines within the box plots display the medians, the upper line indicates the 75th quartile and the lower line, the 25th quartile. The whiskers indicate the maximum and minimum ranges. Each individual specimen is indicated by a dot withing the boxes.

Table 1. Biliverdin (BV) and Total Fluorescent Compound (TFC) data for mullet (*M. liza*, n = 29) specimens sampled from Guanabara Bay, southeastern Brazil

Parameter	BV (µg mL^{-1})	TFC (1-OH-Pyrene equivalents)
Minimum	4.83	0.03
Maximum	63.96	2.05
Mean ± SD	25.11 ± 13.84	0.69 ± 0.53

Biliverdin concentrations ranged from 4.83 to 63.96 and TFC, from 0.03 to 2.05 when grouping all samples regardless of sex, and no significant differences (p > 0.05) were observed when comparing males (BV: 7.33 to 46.89; TFC: 0.03 to 2.05) and females (BV: 4.83 to 63.96; TFC: 0.14 to 1.31) for these biliary evaluations (Figure 3).

The lack of significant differences between sexes for BV and TFC seems to indicate that PAH excretion through the biliary route does not seem to be affected by potential reproduction efforts. This adds to another biliary PAH monitoring advantage, as biliary PAH excretion is already preferable for PAH exposure monitoring efforts in fish as routinely monitored tissues, such as muscle and liver usually present low or undetectable PAH levels, due to rapid metabolization levels (Beyer et al., 2010). Therefore, PAH metabolites reflect fish exposure to PAHs from about 2 to 3 days prior to the analyses, uninfluenced by potential fish migration activities (Ariese et al., 1993).

The BV and TFC results for *M. liza* bile are displayed in Table 2. Data are presented as means ± standard deviation.

Biliverdin concentrations ranged from 4.83 to 63.96 and TFC, from 0.03 to 2.05 when grouping all samples regardless of sex, and no significant differences (p > 0.05) were observed when comparing males (BV: 7.33 to 46.89; TFC: 0.03 to 2.05) and females (BV: 4.83 to 63.96; TFC: 0.14 to 1.31) for these biliary evaluations (Figure 3).

The lack of significant differences between sexes for BV and TFC seems to indicate that PAH excretion through the biliary route does not seem to be affected by potential reproduction efforts. This adds to another biliary PAH monitoring advantage, as biliary PAH excretion is already preferable for PAH exposure monitoring efforts in fish as routinely monitored tissues, such as muscle and liver usually present low or undetectable PAH levels, due to rapid metabolization levels (Beyer et al., 2010). Therefore, PAH metabolites reflect fish exposure to PAHs from about 2 to 3 days prior to the analyses, uninfluenced by potential fish migration activities (Ariese et al., 1993).

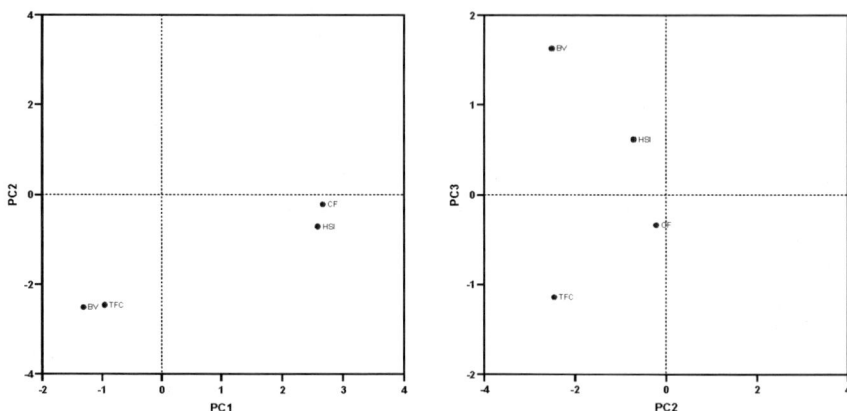

Figure 4. Principal Component Analysis (PCA) indicating the three main Principal Components responsible for the data variations observed for mullet (*M. liza*, n = 29) specimens sampled from Guanabara Bay, southeastern Brazil. Grey dots indicate PC scores and blue ones, PC loadings.

The Principal Component Analysis results regarding BV, TFC, CF and HSI in the sampled mullet specimens are displayed in Table 2, while the PCA biplots for the assessed data are displayed in Figure 4.

The first three principal components (PC) were the most representative, explaining 95.31% of the total data variability (PC1 = 44.18%; PC2 = 34.84%; PC3 = 16.29%). A significant positive association was detected between CF and HSI, which is to be expected as increased HSI indicate reproduction efforts and seemingly healthy individuals.

Table 2. Principal Component Analysis results regarding biliverdin (BV), total fluorescent compounds (TFC), condition factor (CF) and hepatosomatic index (HSI) in mullet (*M. liza*, n = 29) specimens sampled from Guanabara Bay, southeastern Brazil

Principal Component Analysis			
PC	PC1	PC2	PC3
Eigenvalue	1.767	1.393	0.6516
Proportion of variance	44.18%	34.84%	16.29%
Cumulative proportion of variance	44.18%	79.02%	95.31%
Loadings			
	PC1	PC2	PC3
BV	-0.196	-0.825	0.515
TFC	-0.141	-0.809	-0.567
CF	0.939	-0.070	-0.166
HSI	0.909	-0.232	0.194
Eigen vectors			
	PC1	PC2	PC3
BV	-0.148	-0.699	0.638
TFC	-0.106	-0.685	-0.702
CF	0.707	-0.059	-0.206
HSI	0.684	-0.197	0.241
Variable contributions			
	PC1	PC2	PC3
BV	0.022	0.489	0.407
TFC	0.011	0.469	0.493
CF	0.499	0.003	0.042
HSI	0.468	0.039	0.058

A negative association between BV and TFC was noted in mullet bile, which seems to indicate that lower BV amounts, are associated to lower TFC. Under normal conditions, bile, and any contaminants present therein, are concentrated in the gallbladder. For example, during starvation periods, the amounts of PAH sulphated and glucuronidated metabolites in the gallbladder have been reported as increasing significantly (Beyer et al., 1997). When fish feed, the beginning of digestion in the upper part of the gastrointestinal tract leads to gallbladder emptying into the intestine via the common bile duct, followed by quick filling with water, decreasing pollutant concentrations (Richardson et al., 2004). Furthermore, a recent study has also reported that fish feeding status also affects biliary biomarkers of exposure to other contaminants, such as metals (Hauser-Davis et al., 2021a), indicating the importance of this parameters in biomonitoring effects concerning different types of pollutants in aquatic systems. Therefore, the results observed herein corroborate the fish bile pollutant metabolism route reported in the literature.

To a lesser extent, associations were noted between the FCF and TFC and between the HSI and BV when evaluating PC2 and PC3. The former seems to indicate that TFC, due to their teratogenic, and mutagenic potential, may affect the general health status of exposed fish, while the latter may be linked to the biliary production by the liver and storage in the gallbladder, with the amount of produced bile by the liver directly affecting BV concentrations in this biological fluid. This has been previously reported for other contaminants, such as metals. For example, Hauser-Davis et al., (2019) and Hauser-Davis et al., (2021b) demonstrated a moderate negative correlation between intracellular copper and the FCF in a teleost, *Dules auriga*, from two different sampling sites in southeastern Brazil, including GB, indicating that increasing environmental copper concentrations may decrease the general health status of this species. In contrast, other assessments of this type in the literature reported no significant correlations between contaminants and morphometric indices. In this regard, Hauser-Davis et al., (2021b) noted a lack of associations between the FCF and HSI to PFAS, which also display the potential to cause teratogenic, mutagenic effects, immunotoxic, reproductive and/or developmental effects (Lau et al., 2007). Therefore, it seems that differential contamination inputs differentially affect fish morphometric indices, which merits further assessments as to which contaminants are ideally monitored through these indices. Confounding factors, such as nutritional status, diseases or trophic pressures, however, should also be evaluated, as these may obscure any associations (Yang and Baumann, 2006).

Conclusion

The findings reported herein indicate that biliary PAH excretion does not seem to be affected by reproduction efforts, as no significant differences between sexes were noted for either BV and TFC. This demonstrates an advantage concerning the use of bile in biomonitoring efforts, as PAH concentrate in this fluid and do not suffer the influence of biological events. The negative association observed between TFC and BV corroborates literature reports regarding biliary xenobiotic concentration and subsequent body removal through new feeding events. Both the FCF and TFC and the HSI and BV were statistically associated, indicating potential deleterious TFC health effects in mullet from Guanabara Bay, and biliary production by the liver and storage in the gallbladder, directly affecting BV concentrations, respectively. These results indicate that associations between biliary PAH, biomorphometric indices and biliverdin applied as a feeding status proxy in mullet are an interesting alternative to evaluate biliary xenobiotic detoxification and contaminant effects and may be useful in biomonitoring efforts in this scenario, with the possibility of being applied in other aquatic ecosystems worldwide.

References

Aas, E., Baussant, T., Balk, L., Liewenborg, B., Andersen, O. K. (2000). "PAH metabolites in bile, cytochrome P4501A and DNA adducts as environmental risk parameters for chronic oil exposure: a laboratory experiment with Atlantic cod." *Aquatic Toxicology* 51, 241-258. https://doi.org/10.1016/ s0166-445x(00)00108-9.

Aas, E., Beyer, J., et al., (1998). "PAH in fish bile detected by fixed wavelength fluorescence." *Marine Environmental Research*, 46(1-5), 225-228. https://doi.org/10.1016/S0141-1136(97)00034-2.

Abas, R., Mohamad, S. (2011). Hazardous (Organic) Air Pollutants. Encyclopedia of *Environmental Health*, 23-33. https://doi.org/10.1016/B978–0–444–52272–6.00070–2.

Albieri, R. J., Araújo F. G. (2010). "Reproductive biology of the mullet *Mugil liza* (Teleostei: Mugilidae) in a tropical Brazilian bay." *Zoologia*, 27(3): 331-340. https://doi.org/10.1590/S1984-46702010000300003.

Ariese, F., Kok, S. J., Verkaik, M., Gooijer, C., Velthorst, N. H., Hofstraat, J. W. (1993). "Synchronous fluorescence spectrometry of fish bile: a rapid screening method for the biomonitoring of PAH exposure." *Aquatic Toxicology* 26: 273-286. https://doi.org/10.1016/0166–445X(93)90034–X.

Beyer, J., Jonsson, G., Porte, C., Krahn, M. M., Ariese, F. (2010). "Analytical methods for determining metabolites of polycyclic aromatic hydrocarbon (PAH) pollutants in fish bile: A review." *Environmental Toxicology and Pharmacology*. 30(3): 224-44. https://doi.org/10.1016/j.etap.2010.08.004.

Beyer, J., Sandvik, M., Hylland, K., Fjeld, E., Egaas, E., Aas, E., Skåre, J. U., Goksøyr, A. (1996). "Contaminant accumulation and biomarker responses in flounder (*Platichthys flesus* L.) and Atlantic cod (*Gadus morhua* L.) exposed by caging to polluted sediments in Sørflorden, Norway." *Aquatic Toxicology*, 36, 75-98. https://doi.org/10.1016/S0166-445X(96)00798-9.

Boyer, J. L. (2013). "Bile formation and secretion." *Comprehensive Physiology*, 3(3): 1035-1078. https://doi.org/10.1002/cphy.c120027.

Claro, R., Lindeman, K. C., Parenti, L. R. Ecology of the Marine Fishes of Cuba, *Smithsonian Institution*, 2014.

Da Costa, M. R., Martins, R. R. M., Tomás, A. R. G., Tubino, R. A., Monteiro-Neto, C. (2021). "Biological aspects of *Mugil liza* Valenciennes, 1836 in a tropical estuarine bay in the southwestern Atlantic." *Regional Studies in Marine Science*, 43, 101651. https://doi.org/10.1016/j.rsma.2021.101651.

Dallarosa, J. B., Monego, J. G., Teixeira, E. C. Stefens, J. L. Wiegand, F. (2005). "Polycyclic aromatic hydrocarbons in atmospheric particles in the metropolitan area of Porto Alegre, Brazil." *Atmospheric Environment*, 39, 1609-1625. https://doi.org/10.1016/j.atmosenv.2004.10.045.

Esper, M. D. L. P., de Menezes M. S., Esper W. (2000). "Escala de desenvolvimento gonadal e tamanho de primeira maturação de fêmeas de *Mugil platanus* Günther, 1880 da Baía de Paranaguá, Paraná, Brasil." *Acta Biológica Paranaense*, 29. http://dx.doi.org/10.5380/abpr.v29i0.594.

Fähnrich, K. A., Pravda M., Guilbault, G. G. (2002) "Immunochemical detection of Polycyclic Aromatic Hydrocarbons (PAHs)." *Analytical Letters*, 35:8, 1269-1300. https://doi.org/10.1081/AL-120006666.

Fistarol, G. O., Coutinho, F. H., Moreira, A. P., Venas, T., Cánovas, A., de Paula, S. E. Jr., Coutinho, R., de Moura, R. L., Valentin, J. L., Tenenbaum, D. R., Paranhos, R., do Valle, R. de A, Vicente, A. C., Amado Filho, G. M., Pereira, R. C., Kruger, R., Rezende, C. E., Thompson, C. C., Salomon, P. S., Thompson, F. L. (2015). "Environmental and Sanitary Conditions of Guanabara Bay, Rio de Janeiro." *Frontiers in Microbiology*, 6:1232. https://doi.org/10.3389/fmicb.2015.01232.

Hauser-Davis, R. A., Silva-Junior, D. R., Linde-Arias, A. R., Vianna, M. (2019). "Hepatic Metal and metallothionein levels in a potential sentinel teleost, Dules auriga, from a Southeastern Brazilian Estuary." *Bulletin of Environmental Contamination and Toxicology*, 103, 538-543. https://doi.org/ 10.1007/s00128-019-02654-6.

Hauser-Davis, R. A., Silva-Junior, D. R., Linde-Arias, A. R., Vianna, M. (2021). "Cytosolic and Metallothionein-bound hepatic metals and detoxification in a sentinel teleost, Dules auriga, from Southern Rio de Janeiro, Brazil." *Biological Trace Element Research*, 199, 744-752. https://doi.org/10.1007/ s12011-020-02195-8.

Hauser-Davis, R. A., Ziolli, R. L. (2021). "Biliary Fish Proteomics Applied to Environmental Contamination Assessments: A Case Study in Southeastern Brazil." *Bulletin of Environmental Contamination and Toxicology*. (In press). https://doi.org/10.1007/s00128-021-03104-y.

Honda, M., Suzuki, N. (2020). "Toxicities of Polycyclic Aromatic Hydrocarbons for Aquatic Animals." *International Journal of Environmental Research and Public Health*, 17(4): 1363. https://doi.org/10.3390/ijerph17041363.

Kim, K. H., Jahan, S. A., Kabir, E., Brown, R. J. (2013). "A review of airborne polycyclic aromatic hydrocarbons (PAHs) and their human health effects." *Environment International*, 60: 71-80. https://doi.org/10.1016/j.envint.2013.07.019.

Latimer, J. S., J. Zheng. The Sources, Transport, and Fate of PAHs in the Marine Environment. P. E. T. Douben (ed.), *PAHs: An Ecological Perspective*. John Wiley & Sons Incorporated, New York, NY, 9-33, (2003).

Lau, C., K. Anitole, C. Hodes, D. Lai, A. Pfahles-Hutchens, and J. Seed. (2007). "Perfluoroalkyl Acids: A Review of Monitoring and Toxicological Findings." *Toxicological Sciences* 99: 366-394. https://doi.org/10.1093/toxsci/kfm128.

Lemos, V. M., Varela A. S. Jr., Schwingel P. R., Muelbert J. H., Vieira, J. P. (2014). "Migration and reproductive biology of *Mugil liza* (Teleostei: Mugilidae) in south Brazil." *Journal of Fish Biology*, 85(3): pp. 671-687. https://doi.org/10.1111/jfb.12452.

Maddock, D. M., Burton, M. P. (1998). "Gross and histological of ovarian development and related condition changes in American plaice." *Journal of Fish Biology*, 53: 928-944. https://doi.org/10.1111/j.1095-8649.1998.tb00454.x.

Mansour, S. A., Sidky, M. M. (2002). "Ecotoxicological Studies: 3. Heavy Metals Contaminating Water and Fish from Fayoum Governorate, Egypt." *Food Chemistry*, 78, 15-22. http://dx.doi.org/10.1016/S0308–8146(01)00197–2.

McCarthy, T. C., Christopher J. Sinal (2005). Biotransformation, *Encyclopedia of Toxicology*, 1, Elsevier.

Nelson, J. S., 2006. *Fishes of the World*. Hoboken (New Jersey, USA): John Wiley & Sons.

Prestelo, L., Vianna M. (2016). "Identifying multiple-use conflicts prior to marine spatial planning: A case study of a multi-legislative estuary in Brazil." *Marine Policy*, 67, 83-7. https://doi.org/10.1016/j.marpol.2016.02.001.

Richardson, D. M., Gubbins, M. J., Davies, I. M., Moffat, C. F., Pollard, P. M. (2004). "Effects of feeding status on biliary PAH metabolite and biliverdin concentrations in plaice (*Pleuronectes platessa*)." *Environmental Toxicology and Pharmacology*, 17(2): 79-85. https://doi.org/10.1016/j.etap.2004.03.003.

Van der Oost, R., Beyer, J., Vermeulen, N. P. E. (2003). "Fish bioaccumulation and biomarkers in environmental risk assessment: A review." *Environmental Toxicology and Pharmacology*, 13:57-149. https://doi.org/10.1016/S1382-6689(02)00126-6.

Yang, X., Baumann, P. C. (2006). "Biliary PAH metabolites and the hepatosomatic index of brown bullheads from Lake Erie tributaries." *Ecological Indicators*, 6: 567-574. https://doi.org/10.1016/j.ecolind.2005.08.025

Reviewed by Sylvia Nogueira Land, MSc.

Index

A

abiotic stress, 63, 74, 76
acid, 7, 46, 56, 69
activity level, 47
adaptation, 65
adsorption, 8, 10, 28
adverse effects, 62, 65
aerosols, 34
agricultural sector, 64
agriculture, 62
air pollutants, 3, 5, 8, 12, 14
air quality, vii, 1, 2, 12, 19, 21, 26, 27, 30, 31, 34
air temperature, 14
ambient air temperature, 14
anthocyanin, 71, 75
antioxidant, 64, 70, 71, 74
apex, 42, 45
aquatic systems, 80, 106
aromatic compounds, 42, 66
aromatic hydrocarbons, viii, ix, 2, 32, 34, 36, 54, 57, 58, 59, 61, 62, 63, 66, 68, 75, 77, 80, 90, 96, 108
aromatic rings, viii, 2, 38, 39, 46, 61, 96
artificial intelligence, vii, 1, 2, 13, 17, 30
ascorbic acid, 64
atmosphere, vii, 1, 3, 8, 11, 21, 28
atmospheric deposition, 10, 39
atmospheric pressure, 12
attribution, 13

B

bacteria, 10, 34, 69, 70, 71, 74
bacterial strains, 10
barometric pressure, 15
benthic invertebrates, 40
benzene, 2, 66
benzo(a)pyrene, 4, 5, 9, 11, 47, 63, 80, 81, 87
benzo(b)fluoranthene, 12, 13, 63
bile, 41, 46, 47, 96, 97, 99, 100, 102, 104, 105, 106, 107
biliverdin, vii, ix, 96, 97, 98, 100, 103, 105, 107, 109
bioaccumulation, 4, 40, 51, 59, 83, 109
bioavailability, viii, 10, 37, 48, 66
biodegradation, viii, 4, 9, 10, 32, 38, 56, 61, 66, 67, 68, 70, 71, 72
biomarkers, 47, 48, 49, 52, 53, 56, 59, 106, 109
biomass, 12, 65, 66, 71, 73, 96
biomolecules, 64
biomonitoring, ix, 34, 41, 47, 57, 96, 106, 107
bioremediation, viii, 10, 61, 67, 72, 75, 76, 78, 89
biosynthesis, 69, 76
biotic, 64, 65, 68
Brazil, v, vi, vii, ix, 37, 44, 51, 54, 79, 80, 84, 86, 87, 88, 89, 90, 91, 92, 95, 96, 98, 99, 100, 101, 102, 103, 104, 105, 106, 108, 109

Index

C

calibration, 16, 82
cancer, 5, 33, 36, 39, 51, 57, 77, 81
carbon, 9, 31, 40, 74
carbon dioxide, 9, 40
carbon monoxide, 31
carcinogen, 5, 35
carcinogenicity, 3, 4, 10, 39, 51
carotenoids, 64
cartilaginous, 41
cell membranes, 66
changing environment, 65
chemical properties, viii, 37, 46, 63, 83
chlorophyll, 64, 73
chloroplast, 65, 73
chromatography, 15, 16
chronic diseases, 6
clustering, 18, 26, 27
coal, vii, 3, 12, 14, 27, 32
combustion, 2, 3, 12, 19, 20, 27, 28, 38, 47, 62, 73, 96
common bile duct, 106
composition, 36, 64, 67, 77, 92
compost, 9, 10
compounds, viii, ix, 2, 3, 4, 5, 8, 9, 10, 12, 16, 19, 20, 21, 36, 38, 39, 40, 41, 42, 44, 45, 46, 48, 50, 51, 52, 58, 62, 64, 65, 66, 67, 69, 70, 75, 80, 81, 82, 83, 84, 89, 91, 96, 97, 103, 105
conservation, 8, 54, 55, 90
consumption, viii, 7, 38, 42, 51, 54, 59, 80, 81, 87, 88, 91, 92
contaminants, 3, 4, 5, 9, 10, 11, 12, 21, 38, 39, 40, 42, 45, 49, 50, 51, 52, 55, 58, 59, 68, 80, 83, 91, 96, 97, 99, 105, 106, 107
contaminated sites, 8, 10, 32, 33
contaminated soil(s), 5, 7, 9, 35, 67, 69, 75, 76, 77
contaminated water, 71
contamination, viii, ix, 7, 30, 33, 34, 38, 41, 42, 45, 56, 71, 76, 77, 79, 80, 83, 96, 97, 98, 106
correlation(s), 19, 46, 48, 49, 67, 98, 106
crude oil, ix, 14, 39, 49, 50, 54, 68, 73, 79, 91

cultivation, 7, 64
culture, 67, 68
cytochrome, 47, 55, 56, 97, 107
cytoplasm, 65

D

decomposition, 9, 11
defence, 64
deficiencies, 65
deficiency, 63
degradation, 8, 9, 10, 27, 30, 32, 36, 38, 40, 65, 66, 68, 69, 70, 72
deposition, 7, 11, 35, 96
detection, 15, 41, 70, 108
detoxification, 50, 69, 97, 107, 108
diet, 58, 84, 96
diffusion, 8, 28
digestion, 97, 106
diseases, 6, 106
disinfection, 12
dispersion, 21, 27
distribution, vii, 2, 4, 12, 13, 19, 27, 36, 46, 52, 57, 63, 64, 83, 99
DNA, 4, 5, 48, 73, 107
drought, 63, 74

E

ecotoxicological, viii, 37, 38, 42, 53, 57
ecotoxicology, 33, 36, 38, 53, 56, 73, 76, 93
elasmobranchs, vii, viii, 38, 41, 42, 47, 49, 50, 51, 52, 53, 54
emission, 8, 13, 31, 34, 101
emission source(s, 8, 13, 31, 34
energy, 10, 12, 21, 30, 50, 65, 102
environmental conditions, viii, 2, 3, 10, 13, 18, 30, 41, 63
environmental contamination, 50, 53, 96
environmental factors, vii, viii, 2, 13, 64
environmental impact, ix, 53, 79
environmental pollution, 42, 49, 57, 62, 63, 64, 75
environmental protection, 12
environmental stress, 42

enzymatic activity, 47, 50, 52
Europe, vii, 1, 12, 13, 33, 51
evaporation, 21, 28, 33
evidence, 6, 43, 44, 45, 54
excitation, 69, 101
excretion, 40, 96, 97, 103, 104, 106
explainable artificial intelligence, vii, 1, 17, 30
exposure, vii, viii, 2, 5, 6, 29, 32, 34, 35, 38, 39, 40, 41, 42, 43, 44, 46, 47, 49, 50, 51, 52, 53, 54, 55, 58, 59, 63, 68, 74, 76, 80, 81, 84, 86, 89, 97, 103, 104, 106, 107
extinction, 41, 55
extraction, 15, 16, 34, 82
eXtreme Gradient Boosting, 13, 17

F

fish, viii, ix, 6, 7, 38, 40, 41, 52, 53, 54, 55, 56, 58, 59, 80, 81, 82, 83, 84, 85, 86, 87, 88, 92, 96, 97, 98, 99, 101, 102, 103, 104, 105, 106, 107, 108, 109
Fish and Wildlife Service, 55
fisheries, ix, 51, 80, 81, 83, 84, 85, 87, 88, 99
flavonoids, 65, 70
flavonol, 69, 72
fluid, 106, 107
fluorescence, 3, 15, 31, 46, 55, 73, 89, 107
fluorine, 63, 84, 86
food, 5, 6, 7, 14, 30, 36, 40, 41, 51, 53, 59, 64, 70, 76, 89, 92
food chain, 6, 40, 41, 51, 70
food products, 7, 30
food safety, 53
forage crops, 64
force, 17, 18, 26, 29
forest fire, 7, 38, 96
formation, 4, 6, 10, 30, 36, 48, 50, 65, 108
freshwater, vii, ix, 36, 57, 61, 68, 73, 78, 93
fuzzy clustering, 18, 26, 27

G

gallbladder, 41, 105, 106, 107
gastrointestinal tract, 106
gene expression, 69, 72
genes, 64, 69
genome, 64
germination, 65, 66
global economy, 62
glutathione, 47, 64, 75, 76, 97
groundwater, viii, 61
growth, 10, 35, 42, 50, 63, 65, 66, 68, 69, 71, 73, 75, 77, 97
Gulf of Mexico, 44, 45, 46, 55, 80, 88, 91

H

habitat, 55, 64, 68, 84
hazards, 30, 33, 55
health, vii, ix, 4, 5, 12, 30, 31, 32, 34, 48, 49, 50, 51, 52, 53, 62, 64, 80, 90, 97, 106, 107
health effects, 5, 97, 107
health risks, 12, 52, 80
heavy metals, 34, 56
hormone(s), 5, 69, 77
human, vii, ix, 1, 3, 4, 5, 6, 12, 13, 19, 30, 32, 33, 34, 36, 42, 51, 53, 54, 58, 59, 61, 62, 76, 77, 80, 81, 88, 93, 108
human exposure, vii, 1, 3, 5, 6, 12, 54, 59
human health, ix, 4, 5, 30, 33, 34, 36, 51, 53, 58, 61, 62, 80, 93, 108
humidity, 12, 14, 15, 19, 21, 28, 30, 35
hydrocarbons, viii, 61, 62, 66, 68, 71, 73, 75, 76, 84, 90, 98

I

imbalances, 50, 51
incomplete combustion, viii, 2, 7, 61, 62
individuals, 46, 49, 97, 102, 105
indoor environment, 12, 19, 21, 28
indoor environment XE "indoor environment" 3, 12, 19, 21, 28
induction, 46, 47, 48, 50, 51, 52, 56
industrial emissions, 7, 12

industrial processing, 38
industrial wastes, 56
industries, 47, 98
industry, vii, 21, 38, 81, 86, 91
infrared spectroscopy, 15, 31
ingestion, 5, 8, 11, 40
inhibition, 20, 21, 26, 30, 50, 65, 71
interaction effect, 18
invertebrates, 40, 55, 99
ions, viii, 2, 13, 14, 17, 19, 20, 82
isomers, 27, 30
isothermal heating, 16

K

kerosene, 38
kidneys, 52
kinetics, 58

L

light, 3, 10, 13, 63, 68
lipid peroxidation, 48, 50, 53
liquid chromatography, 82
liver, ix, 42, 44, 46, 48, 53, 96, 98, 99, 102, 103, 104, 106, 107
living environment, 65
logic rules, 17
low temperatures, 2, 39, 65

M

machine learning, vii, 1, 13, 19, 30
mangroves, ix, 79, 98
marine, viii, ix, 3, 37, 38, 39, 40, 41, 45, 54, 55, 56, 57, 58, 59, 75, 78, 79, 82, 84, 89, 90, 91, 92, 93, 96, 98, 107, 108, 109
marine environment, viii, 37, 39, 40, 41, 56, 96
marine fish, 54, 56
marine sanctuary, 90
marine species, 3
mass, 15, 16, 21, 31, 32, 90
Mediterranean, 44, 45, 59
membranes, 4, 46

metabolism, 42, 46, 47, 48, 50, 52, 57, 65, 66, 74, 83, 84, 106
metabolites, ix, 4, 5, 6, 41, 42, 46, 47, 48, 49, 65, 70, 76, 96, 97, 104, 105, 107, 109
metal oxides, 11
metals, 10, 42, 62, 91, 106, 108
meteorological conditions, 11, 13, 31, 35
microorganisms, 9, 55, 66, 68, 70
microsomes, 47, 65
microstructure, 28
models, 13, 31, 35, 39, 57
moisture, 9, 10, 28, 63
moisture content, 28, 63
molecular mass, 39
molecular weight, 2, 3, 4, 8, 10, 11, 30, 39, 40, 45, 46, 63, 66, 70, 83
molecules, 40, 49, 62, 65, 70, 97
mollusks, 80, 84, 87
morphological abnormalities, 56
morphometric, 97, 102, 106
municipal solid waste, 3
mutant, 69
mutation, 5

N

naphthalene, 4, 5, 16, 38, 39, 44, 45, 46, 63, 66, 67, 68, 69, 75, 82, 84, 86
National Academy of Sciences, 92
National Oceanic and Atmospheric Administration, 92
natural gas, 14, 34
negative effects, 65
neonates, 44
neurotoxicity, 51
nitrogen, 16, 31, 75
nitrogen dioxide, 31
North America, 13, 49
nutrient(s), 10, 63, 68
nutrition, 6, 76
nutritional status, 106

O

oil, vii, ix, 3, 5, 7, 10, 14, 21, 34, 38, 39, 45, 47, 49, 50, 51, 56, 58, 62, 66, 68, 79, 80, 81, 83, 84, 86, 87, 88, 89, 90, 91, 92, 96, 98, 107
oil spill, ix, 34, 39, 47, 49, 56, 58, 62, 66, 79, 80, 81, 83, 84, 86, 87, 88, 89, 90, 91, 92, 96
operations, 8, 27
organ, 42, 49
organic chemicals, 57
organic compounds, viii, 37, 38, 61, 63, 65, 77, 96
organic pollutants, vii, viii, 2, 9, 13, 38, 54, 61, 62, 65, 66, 70, 96
organic solvents, viii, 61
outdoor environment, vii, viii, 2, 12, 13, 17, 18, 19, 29, 34
oxidation, 8, 9, 33, 97
oxidative damage, 71, 75
oxidative stress, 48, 57, 64
oxygen, 10, 40, 41, 68
ozone, 4, 11, 12, 31

P

perylene, 4, 16, 45, 46, 63, 81, 82, 86
petroleum, viii, 3, 8, 38, 53, 58, 61, 62, 68, 71, 73, 75, 90, 98
pH, 10, 63
phenolic compounds, 64, 75
photooxidative degradation, 27
photosynthesis, 65, 69, 71, 72, 75
photosynthetic performance, 67, 72
phytoremediation, vii, viii, 9, 61, 62, 63, 66, 68, 70, 71, 73, 74, 75, 76, 77
plant growth, 63, 65
plant physiology, 62, 75
plants, vii, viii, 3, 8, 10, 14, 27, 61, 62, 63, 64, 65, 66, 68, 70, 72, 73, 74, 75, 76
PM-bound pollutants, 2, 11, 13, 16
pollutant(s), vii, viii, 2, 4, 6, 9, 11, 13, 14, 17, 19, 21, 26, 27, 28, 30, 38, 39, 42, 45, 51, 53, 54, 56, 59, 61, 62, 63, 64, 65, 66, 68, 70, 76, 96, 97, 106, 107
pollution, viii, 6, 7, 27, 28, 30, 31, 32, 33, 34, 35, 42, 45, 49, 53, 54, 57, 58, 59, 61, 62, 63, 66, 70, 75, 89, 90, 91, 92, 93, 98, 99
polychlorinated biphenyls (PCBs), 47, 54, 58
polycyclic aromatic hydrocarbons (PAHs), v, vi, vii, viii, ix, 1, 2, 3, 4, 5, 6, 7, 8, 10, 11, 12, 14, 15, 16, 17, 19, 21, 26, 28, 29, 31, 32, 33, 34, 35, 36, 37, 38, 39, 40, 41, 42, 44, 45, 46, 47, 48, 49, 50, 51, 52, 53, 54, 55, 56, 57, 58, 59, 61, 62, 63, 65, 66, 67, 68, 69, 70, 71, 72, 73, 74, 75, 76, 77, 79, 80, 81, 82, 83, 84, 85, 86, 87, 88, 89, 90, 91, 92, 95, 96, 104, 108, 109
population, 5, 6, 10, 42, 48, 50, 52, 62, 64, 66, 81, 87, 98
protection, 31, 65, 68, 98
proteins, 64
public health, viii, 38, 52, 53
purity, 100, 101
pyrolysis, viii, 61, 62

Q

quality assurance, 15
quality control, 15, 16
quantification, 17, 44, 46, 82
quartile, 102, 103
quartz, 14, 15, 101

R

radiation, 12, 35, 40, 75, 77
radon, viii, 2, 13, 15, 30, 35
reactions, 4, 8, 97
reactive oxygen, 50, 64
recovery, 16, 17, 74, 82
remediation, 10, 53, 68, 71, 75
reproduction, 40, 50, 98, 103, 104, 105, 106
residues, ix, 36, 54, 79, 88, 93
resistance, 3, 4, 38
resolution, 14, 15
resources, vii, ix, 62, 68, 90
response, 49, 69, 71, 74, 82, 91, 92, 98

risk assessment, 3, 32, 34, 35, 36, 42, 58, 59, 80, 81, 93, 109
root system, 63, 66

S

saturated hydrocarbons, 66
seafood, 6, 32, 35, 52, 80, 81, 87, 88, 90, 91, 92
secondary metabolism, 64
SHapley Additive exPlanations, 13, 18
signal transduction, 64
species, ix, 2, 3, 4, 5, 9, 10, 11, 12, 20, 34, 40, 41, 42, 45, 46, 47, 48, 49, 50, 52, 53, 54, 55, 56, 58, 62, 64, 65, 67, 68, 69, 73, 75, 77, 84, 97, 98, 99, 106
spectroscopy, 31, 89
stability, 68, 70
standard deviation, 86, 101, 102, 104
storage, 40, 42, 62, 106, 107
stress, 41, 56, 62, 63, 65, 68, 70, 71, 74, 76, 77
stress factors, 65
stress response, 69, 70, 76
structure, 2, 3, 8, 10, 66
surface area, 7, 10
surface layer, 28
survival, 42, 50, 51, 53
synthesis, 64, 65

T

temperature, 2, 9, 10, 12, 15, 16, 20, 21, 24, 27, 30, 33, 41, 63, 75, 82, 98
thermal decomposition, 2
tissue, 10, 40, 42, 44, 45, 46, 52, 55, 57, 65, 83, 84
tocopherols, 64
toxic effect, 51, 80
toxicity, 3, 4, 5, 10, 33, 34, 40, 51, 54, 57, 63, 72
trace elements, viii, 2, 13, 14, 15, 16, 17, 19
transport, 9, 10, 11, 14, 19, 21, 33, 39, 69
transportation, vii, 1, 12, 62
treatment, 9, 69, 73

V

vegetables, 6, 7, 34, 71
vegetation, 3, 8, 11
vertebrates, ix, 40, 47, 51, 52, 55, 96

W

waste, 8, 38, 66
waste incineration, 38
wastewater, 66, 98
water, vii, 1, 3, 4, 7, 8, 9, 10, 13, 16, 38, 39, 40, 41, 44, 54, 63, 65, 68, 70, 74, 75, 80, 84, 91, 97, 98, 99, 106
water quality, 68
water resources, 68
wetlands, 71, 78
wildlife, 55
wind speed, 12
wood, vii, 3, 7, 8, 12, 14, 27, 32
worldwide, viii, 42, 43, 53, 61, 63, 99, 107